A Student Workbook to Accompany

COMPANY OFFICER

2ND EDITION

A Student Workbook to Accompany

COMPANY OFFICER

2ND EDITION

CLINTON H. SMOKE

THOMSON

DELMAR LEARNING

Australia Canada Mexico Singapore Spain United Kingdom United States

THOMSON

DELMAR LEARNING

A Student Workbook to Accompany Company Officer, 2nd Edition
Clinton H. Smoke

Vice President, Technology and Trades SBU:
Alar Elken

Editorial Director:
Sandy Clark

Acquisitions Editor:
Alison Weintraub

Development Editor:
Jennifer A. Thompson

Marketing Director:
Dave Garza

Channel Manager:
Bill Lawrensen

Marketing Coordinator:
Mark Pierro

Production Director:
Mary Ellen Black

Production Editor:
Ruth Fisher
Toni Hansen

Editorial Assistant:
Jenessa Cox

For more information contact
Thomson Delmar Learning
Executive Woods
5 Maxwell Drive, PO Box 8007,
Clifton Park, NY 12065-8007
Or find us on the World Wide Web
at *www.delmarlearning.com*

Library of Congress
Cataloging-in-Publication
Data: 2004026477

ISBN: 1-4018-2606-7

On an average day,
the fire service of the United States
responds to 4,300 fire alarms.
These fires kill 11 Americans and
injure 50 others. The property loss is $24 million.
Some days are even worse.

Are you ready to help?

Dunken

CONTENTS

PREFACE

This Student Workbook can assist you in three ways: it is a study companion to the *Company Officer,* 2nd Edition textbook, it will help you meet the requirements for certification as a fire officer and, most important, it can help you become a more effective company officer.

As a companion to the *Company Officer* textbook, this workbook contains information and supporting material for each of the thirteen chapters in the textbook. Specifically, it contains the learning objectives and offers a brief outline of each chapter so that you can quickly see an overview of the material and add your personal notes regarding its content.

Selected additional material is presented. Some of this material may be repeated from the text for emphasis. Some of the material may be new—a slightly different way of presenting something that was covered in the text. Both are offered to help your understanding. Discussion questions are presented so you can answer the questions directly in this workbook.

For this second edition, three things were added. First, the materials now address the requirements for Fire Officer II as well as Fire Officer I. Second, we added a quiz consisting of multiple-choice questions. These provide a good check of your mastery of the subject. And finally, we added student activities that are specifically designed to satisfy the requirements for Fire Officer I and II.

For those seeking certification, we suggest that you start by becoming familiar with the requirements for certification as a Fire Officer I and II as presented in NFPA 1021, *Standard on Fire Officer Professional Qualifications* 2003 Edition. Thanks to the National Fire Prevention Association (NFPA), that document appears in the textbook as Appendix A.

In addition to the information required for certification, useful information and practical applications are emphasized to assist you in mastering the material. The information and practical applications will help you understand the book, pass a class, gain certification and, most important, help you become a more effective officer.

What Is In a Word?

In some cases, firefighter terminology seems to be favoring the career fire service while excluding the volunteer fire service. In reality, that is not the case. Most of the nation's fire service is made up of volunteer fire departments.

Although NFPA 1021 is entitled "Standard for Fire Officer Professional Qualifications," its guidelines are equally applicable to both career organizations and volunteer departments. In addition, career development programs, despite their professional sounding name, are valuable in any organization, including both career and volunteer fire departments. The term *personal development* may be more appropriate.

Mastering the Material

For those using these materials as a formal training program or college class, we suggest the following actions to help you learn the material:

1. Read the table of contents to get an overview of the material to be covered.
2. Read one chapter at a time. Start by reading the objectives for the chapter, and then read the text material in the chapter.
3. While reading, feel free to highlight items and make notes. If something is not clear, make a note of that, too. Maybe it will be cleared up in subsequent paragraphs. If not, you may want to ask your instructor for additional information.
4. You may want to keep the appropriate pages of the Student Workbook before you as you read the text.
5. While reading, note the key terms and important concepts that are highlighted in the book.
6. Answer the review questions at the end of the text. Just looking up the answers will help you fill in the blanks but will not aid in your long-term learning of the material. You should study the material until you are able to answer the questions without looking at the text. Once you have answered the question, you should certainly look back and check your answer.

7. At this point, you should be ready to participate in class. The classroom experience should provide you with an opportunity to learn more; it should not be an oral presentation of the contents of the textbook. Hopefully, you will have time to consider the review and discussion questions during the class discussions.

8. After class, try to find time to reread the text. For many students, this is the part of the learning process they are most likely to omit. Ironically, it is the most important part. While reading the material a second time, you may want to take time to outline the information.

9. By now you have read the material twice and covered it in class as well. You should be comfortable with everything covered in the chapter. Before moving on to the next chapter, ask yourself, "Do I completely understand this material?"

10. Each chapter includes a list of objectives; satisfy yourself that you have learned each objective.

11. Each chapter contains a series of review and discussion questions. When you finish each chapter, you should be able to answer each question without looking at the text or your notes.

12. In addition to the questions in the textbook, the Student Workbook contains activities that will further assist you in satisfying the requirements of NFPA 1021, Fire Officer Professional certification.

INTRODUCTION

About the Book's Organization

The textbook is designed to help you gain competency in the requirements of NFPA 1021. That standard lists the requirements for Fire Officer I and II under six broad categories: administration, community and government relations, human resource management, inspection and investigation, emergency service delivery, and safety.

The textbook is organized into chapters covering those major topics and several other topics that are addressed as prerequisite knowledge items in NFPA 1021. The first several chapters deal with communications, organizations, management, and leadership. These tools are then applied to the process of preparing for and managing emergency incidents (see **Figure P-1**).

Chapter 1 starts with a discussion of the vital role of the company officer, examines the challenges that face company officers, and outlines reasons for being well-prepared when they move into this important position.

Chapter 2 is about communications. This topic is so important that it is placed at the front of the book. It is a tool you will be using throughout the remainder of the course, as well as in your work and personal life.

Chapter 3 addresses the concept of organizations and the company officer's role in them. Understanding your organization helps define your place and purpose in it. This chapter also sets the stage for the chapters on management and leadership that follow.

Chapter 4 briefly reviews the field of management science and examines the company officer's role as a manager.

Chapter 5 provides some guidelines for practical applications of modern management principles at the company level. This chapter also discusses labor relations and customer service in the fire service.

Chapter 6 briefly examines leadership principles and the company officer's role as a leader. Two important topics are discussed in this chapter: leading a diverse work force and the prevention and management of sexual harassment in the workplace.

Chapter 7 offers some effective tools for making leadership work at the company level.

Chapter 8 addresses the important role of the company officer in administering the department's occupational safety and health programs at the company level.

Chapter 9 deals with fire prevention. Civilian injuries, fatalities, and property loss due to fire should be a concern to company officers. Supporting efforts to prevent these losses from occurring in your community is a part of your job.

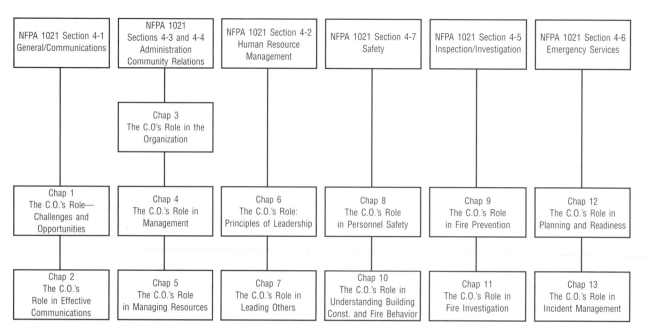

Figure P-1 Comparison of the requirements of Chapter 4 of NFPA 1021 and the Chapters in *Company Officer*. Chapter 4 addresses the requirements for Fire Officer I. Chapter 5 addresses the same topics at the Fire Officer II level.

Chapter 10 reviews building construction and how fire destroys buildings. By better understanding building construction and fire behavior, you can be more effective in dealing with the threat of fire.

Chapter 11 provides a brief overview of fire-cause determination, so that you can accurately determine the cause of fires, especially when arson is indicated.

Chapter 12 addresses your responsibilities for keeping yourself and your company trained, fit, and ready to deal with expected emergencies. For most organizations, "waiting for the big one" is no longer an acceptable activity. Fire service personnel are expected to use every waking moment to prepare for events and to take action to reduce their consequences.

Finally, Chapter 13 deals with resource management at the scene of an emergency.

Some students suggest that allocating only one chapter to emergency management is not enough, and that allocating five chapters to organization, management, and leadership is too much. Most experienced fire officers will tell you that they are adequately prepared to handle the emergency situations they encounter. Many will admit that they are not adequately prepared for the administrative problems they face. Most would be happy to trade the experience they gained by trial and error for a good training program that addresses the administrative issues along with the emergency situations.

Reading This Book as Part of a Certification Program

If you are reading this material as part of an officer certification program or training course, you may be asked to demonstrate both oral and written communications skills. This is to help you become more effective in both of these areas.

During this course, you may also be asked to prepare written homework. The homework will consist of several short assignments that require a written response. The types of responses represent the various forms of written communications you can expect to prepare as an officer. You may also be called on to make an oral report in class.

If you are called upon to speak or write, your efforts will be evaluated, and more important, you will be coached on ways to be more effective. While not every instructor of this course is able to provide you with the same type of feedback that a college English or speech professor might be able to offer, most instructors can look at a letter or report, or listen to you speak, and assess your communications skills. The instructor's purpose is not so much to assign a grade as to determine whether or not you were effective and help you become even more so.

In the case of written communications, an instructor should be able to determine if your reply is neat, answers the questions, and reads well.

Your mastery of other material in this course can be demonstrated in several ways. A traditional approach is to take a test. An alternate approach is to ask you to answer all of the questions and do all of the exercises in this Student Workbook to a satisfactory level. The manual covers all of the elements of NFPA 1021 for Fire Officer I and II.

To validate your efforts, the authority providing your certification may ask for further evidence of your personal competence by asking you to take an oral or written test. If you have seriously studied the material and answered all of the questions, you should be completely ready for that challenge (see **Figure P-2**).

Another alternative is to ask you to demonstrate your knowledge as well as your oral and written communications skills in a variety of practical situations. To provide a reasonable assessment environment, you should be allowed to work with the tools you would normally use if writing. These tools should include a computer, assorted references on writing, and the luxury of a little time to organize, prepare your ideas, and polish your prose.

One Final Comment

While the book was written to address the requirements for Fire Officer I and II as listed in NFPA 1021, we tried to make the material both interesting and useful for both volunteers and paid personnel who operate in all emergency service organizations. The issues for supervisors, administrators, and managers in rescue squads and Emergency Medical Service (EMS) organizations are very similar to those found in the fire service. When possible, the textbook and this manual address the needs of these individuals as well.

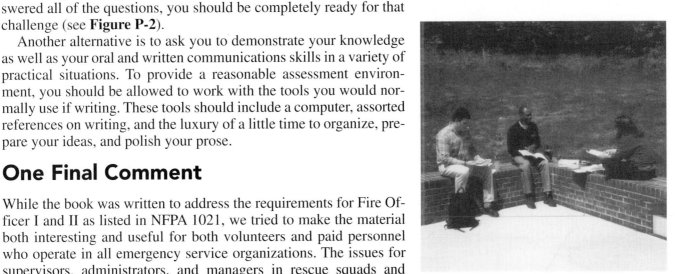

Figure P-2 These firefighters and students are preparing for a test as part of their officer training program.

THE COMPANY OFFICER'S ROLE—CHALLENGES AND OPPORTUNITIES

 ## OUTLINE

- Objectives
- Overview
- The Many Roles of the Company Officer
- Professional Development: An Opportunity for Company Officers

- Strategies for Success as a Company Officer
- Key Terms
- Review Questions

- Discussion Questions
- Homework
- Chapter Quiz

 ## OBJECTIVES

Upon completion of this chapter, you should be able to describe:
- The duties of a company officer
- The national standard for competency for fire officers
- Professional development opportunities for company officers
- Strategies for success as a company officer

OVERVIEW

Being an officer in the fire service is one of the greatest jobs on earth. At times, you will be your own boss. At other times, you will supervise others. Often, you will be doing many of the things you came into the fire service to do without having to do some of the dirty work. Being a company officer is not easy, but it has its rewards. One of those rewards is working with others, often as a leader, to do the very things that motivated you to enter this profession in the first place—saving lives and property. There is no higher calling (see **Figure 1-1**).

There are significant rewards associated with the position of company officer, but there are also significant responsibilities. Our purpose here is to help you prepare to meet these responsibilities. In this section, you will find useful professional information from two large, progressive fire departments. They both have well-established documentation to help their personnel connect with their department's promotion program. This advice is useful everywhere.

Figure 1-1 Fire officers are leaders *and* team members. As a team, they face new challenges every day. Here the team carries a patient suffering from injuries resulting from a fall while rock climbing in a national park. Photo courtesy of Fairfax County Fire and Rescue Department.

THE MANY ROLES OF THE COMPANY OFFICER

Coach	Innovator	Public relations representative
Communicator	Instructor	Referee
Counselor	Leader	Role model
Decision-maker	Listener	Safety officer
Evaluator	Manager	Student
Firefighter	Motivator	Supervisor
Friend	Planner	Writer

As shown in the accompanying list, you are expected to be able to perform many roles. You must learn how to perform these roles while retaining your previous skills and knowledge. Company officers do not normally have the luxury of standing around at the scene of emergencies with a radio in one hand, directing the other fire-fighters' activities. They frequently help fight fires, or help diagnose and treat patients, along with the rest of the troops.

With the minimal staffing of companies seen in many fire departments, you, as the company officer, are expected to be a firefighter along with the rest of the company. In addition, you are expected to manage and supervise your subordinates. Many argue that the job of company officer is one of the most demanding in the organization.

 # PROFESSIONAL DEVELOPMENT: AN OPPORTUNITY FOR COMPANY OFFICERS

Introduction

The fire service is unique. No other industry serves the public in so many ways. The ability to react to the public's needs and resolve any problems encountered places a high level of responsibility on every member of the department, including the newest firefighter. For these reasons, demanding entry-level requirements and rigorous recruit training are vitally important. Because of the complex nature of the profession, firefighters are carefully screened before entering the department, and once in the department, all members should be continuously trained to help them maintain a high level of proficiency.

In order to maintain these skills and learn new ones, a continuing program of training and education should be provided to all department members. This training helps members maintain proficiency at their present level, meet certification requirements, learn new procedures, and become familiar with new technologies. This program will help members develop the skills, knowledge, and abilities needed for promotion and successful service in positions of increasing responsibility.

 # STRATEGIES FOR SUCCESS AS A COMPANY OFFICER

Be a professional. Being a professional means that you are dedicated and committed to the job, learning all that you can, and giving as much as you can. Work to make yourself, your company, and your department the very best possible, whether you are paid or volunteer your services.

Set personal goals. Every time you go to the station, make some contribution to the improvement of the place and your organization.

Continuously work on your own training and education, and encourage the same in others. Training is a part of the job, so obtain all of the training and education that you can. Set personal training and education goals, and use your time and energy to attain them.

Be loyal to your colleagues and your department. Speak well of your job, your department, and your associates. Although the Constitution guarantees freedom of speech, good judgment suggests that this freedom must be tempered with reason and, sometimes, with constraint.

Be a role model. You will likely be able to recall the first officer or first instructor you had as a fire-fighter, Emergency Medical Technician (EMT), or rescue squad member. These individuals leave a lasting impression upon us—hopefully positive. For those who are in your charge, the same holds true; be a role model for them. Help them develop a professional attitude about the job and be all they can be. Being a good role model does not mean you are perfect. It means you try hard and work to improve so you can do better tomorrow. Being a role model means being a professional (see **Figure 1-2**). Professionalism encompasses attitude, behavior, communication style, demeanor, and ethical beliefs. It is as simple as **A, B, C, D,** and **E.**

Attitude is at the core of your performance. A good positive attitude suggests that you are working to be a role model.

Behavior is how you act. You are watched by others while you are on duty and when you are off duty. Your actions reflect on you, your department, and your profession.

Figure 1-2 The company officer should serve as a role model for others.

Communication is how you get your ideas across to others. Communication may be oral, written, or nonverbal. We are in a "people" business; we work with people, and we serve people. Today's emergency service organizations spend a great deal of time working to improve our human relations side so that we can better work together and serve our citizens.

Demeanor embraces all three of the previously listed items. Focus your energies on the mission of your organization, and use positive attitude, good personal behavior, and effective communications to accomplish the goals of your organization.

Ethics deal with conforming to the highest professional standards of your organization. It is more than just words, and it is more than just acting out words. It is doing the right thing, every time, every day.

Preparing for Promotion

While routine training is provided as part of the member's job, preparing for advancement will require considerable individual initiative on the member's part. Advancement brings greater prestige and increased pay. It also brings added responsibilities. The Professional Development Program requires that new skills, knowledge, and abilities be mastered before the member becomes eligible for promotion to the next rank. The requirements in the

Professional Development Program should be the result of long and careful consideration and based on the requirements of the position as well as the minimum standards set forth by the National Fire Protection Association and the state agency that regulates fire officer certification.

The standards for promotion to the supervisory ranks are demanding. These standards reflect the significant requirements placed on supervisory personnel while performing their duties. The primary purpose of the standards is to ensure that personnel are fully prepared to face the complex challenges of their positions. Most of these standards focus on administration, management, and supervisory issues.

We have mentioned the standard for fire officers. Let's look at its origins.[1]

In 1971, the Joint Council of National Fire Service Organizations (JCNFSO) created the National Professional Qualifications Board (NPQB) for the fire service to facilitate the development of nationally applicable performance standards for uniformed fire service personnel. On December 14, 1972, the Board established four technical committees to develop those standards using the National Fire Protection Association (NFPA) standards-making system. The initial committees addressed the following career areas: firefighter, fire officer, fire service instructors, and fire inspector and investigator. In July 1976, the Association adopted the first edition of NFPA 1021.

The original concept of the professional qualification standards, as directed by the JCNFSO and the NPQB, was to develop an interrelated set of performance standards specifically for the fire service. The various levels of achievement in the standards were to build on each other within a strictly defined career ladder. In the late 1980s, revisions of the standards recognized that the documents should stand on their own merit in terms of job performance requirements for a given field. Accordingly, the strict career ladder concept was abandoned, except for the progression from firefighter to fire officer. The later revisions, therefore, facilitated the use of the documents by other than the uniformed fire services.

In 1990, the responsibility for the appointment of professional qualifications committees and the development of the professional qualifications standards were assumed by the NFPA. The 1992 edition of NFPA 1021 reduced the number of levels of progression in the standard to four. In the 1997 edition, NFPA 1021 was converted to the job performance requirement (JPR) format to be consistent with the other standards in the Professional Qualifications Project. Each JPR consists of the task to be performed; the tools, equipment, or materials that must be provided to successfully complete the task; evaluation parameters and/or performance outcomes; and lists of prerequisite knowledge and skills one must have to perform the task.

The intent of the technical committee was to develop clear and concise job performance requirements that could be used to determine that an individual, when measured to the standards, possesses the skills and knowledge to perform as a fire officer. The committee further contends that these job performance requirements can be used in any fire department in any city, town, or private organization throughout North America.

In preparing the 2003 edition of the document, the technical committee did a task analysis to validate the continued need for the use of four levels in the document. It was found that several tasks were actually being performed at a level lower than indicated in the previous edition. Changes were made to reflect that fact, as well as to bring the document into conformance with the new NFPA Manual of Style.

What Is a Company Officer?

The term **company officer** includes all of those in a supervisory capacity at the company level. Titles such as lieutenant and captain are commonly used.

A **lieutenant,** under the direction of a senior officer, serves as a commander of a functional unit such as a truck company or rescue squad within a multifunction station. In the absence of a more senior officer, the lieutenant may assume command of an entire shift. As such, the lieutenant is required to assume command at the emergency scene, pending the arrival of a senior officer. Lieutenants also conduct training, make assignments, prepare reports, maintain records, and make recommendations including those regarding the performance of subordinates. Lieutenants may be assigned comparable duties in the other divisions of the department.

A **captain** typically serves as shift commander in a station. As such, the captain must plan and assign work for all shift personnel. During emergency operations, the captain must be able to evaluate fire and rescue emergencies, determine the necessity for additional resources as well as the proper course of action, assume command at the scene pending the arrival of a senior officer, and assist in mitigating the emergency situation. A captain will also conduct training, prepare reports, maintain records, and make recommendations including those regarding the performance of subordinates. They may be assigned comparable duties in the other departmental divisions.

Some departments have established a grade of senior captain, the senior officer at a particular station. A **senior captain** is responsible for overall station management and is directly responsible for all personnel and operations on an assigned shift. The senior captain plans and executes work assignments and manages the station's resources and maintenance needs. During emergency operations, the senior captain shall evaluate fire and rescue emergencies and determine the necessity for additional resources as well as the proper course of action, assume command at the scene pending the arrival of a senior officer, and assist in mitigating the emergency situation. The senior captain also typically coordinates citizen education and information programs and investigates complaints.

Most departments use two of the three titles to designate the positions of company officers. In general, the requirements for these positions align with requirements for Fire Officer I and Fire Officer II in NFPA 1021.

The following information was adopted from the Phoenix Fire Department's Career Development Handbook.

Start Early on Your Professional Development

A successful and expanding profession is the result of careful planning and hard work. Members who are interested in advancement should begin planning a course of action early in their professional lives. Developing a personal action plan can help identify goals; focusing on these goals will reduce the time and effort required to attain them.

Personal professional goals should be established early. Ultimate professional goals could be as high as each individual aspires. It is also important to establish a series of intermediate goals that are both realistic and attainable. These intermediate goals provide benchmarks for measuring progress along the professional path while you keep sight of the ultimate goal (see **Figure 1-3**).

Proper guidance and counseling will assist a member interested in professional development. People with successful professional lives may have suggestions to offer and can serve as role models for one seeking to follow in their footsteps. Role models should be carefully selected based on a record of proven performance and their overall understanding of the department.

Many members limit their growth potential by failing to understand the overall department as an organization. Each member should thoroughly investigate what we do, how we do it, and who is responsible for getting it done. While the answer to these questions may be obvious to some, many members fail to understand how each division of the department contributes to the overall mission of the organization.

In the final analysis, professional success should be judged by performance, not by position. If you do the best that you can at whatever level you choose, you will be professionally successful. (See **Figure 1-4**).

Employee Development Training Programs

The department and the city are actively involved in employee development. Through in-service training, member development programs, and tuition assistance, it is possible for members to improve their knowledge, skills, and abilities at little personal cost. Professional development benefits the community, the department, and the member.

The city offers various training programs for all members. In most cases, the courses are free. The following courses are considered especially useful:

▌ General Development
▌ Supervisory Development
▌ Management Development
▌ Workplace Skills
▌ Workplace Wellness
▌ Computer Training
▌ Quality and Productivity
▌ Secretarial and Office Skills

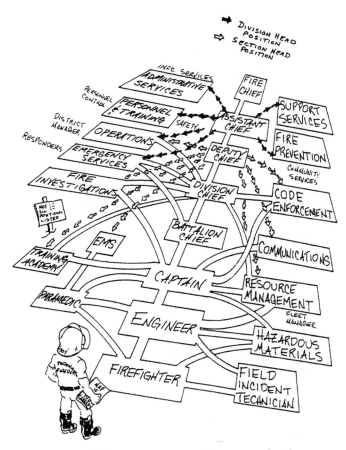

Figure 1-3 There are many options open for those entering the fire service today. Chart courtesy of the Phoenix Fire Department.

Figure 1-4 Working with others is an important part of the company officer's job.

Formal Education

Formal education is essential to professional development and is expected to become even more critical to the promotion process in the future. With careful planning, educational programs can be tailored to meet current needs while also meeting long-term goals. Based on personal professional plans, members should develop individual educational plans that respond to both short-term and long-range goals.

College courses in fire science provide the technical base of knowledge beneficial to all firefighters. These courses are job related and should be considered as immediate goals in your education plan. Fire science courses will prove invaluable throughout your professional experience, but they are of primary importance in the beginning steps of the professional ladder. An associates' degree in fire science is highly recommended for company officers.

A baccalaureate degree provides a well-rounded education program that contributes to personal growth and development. Some degree programs may prove more beneficial than others in terms of job relatedness but, regardless of the program, the education process is extremely valuable. A baccalaureate degree is highly recommended for officers aspiring to move into management positions beyond the company officer rank.

Get Involved

Everything that is accomplished by the department is the direct result of people striving to do their work. Whether long-range planning or carrying out the day's assignments, the individual contributions of members make it happen. You can have a great influence on the direction of the department if you are willing to contribute. All members are urged to seek responsibilities and share their experience and knowledge to improve the department.

Promotion to Captain

The captain is the first level of supervision. Captains are responsible for carrying out organizational objectives. Captains must ensure that their individual units are able to provide timely and quality service to the public in fire suppression, fire prevention, EMS, and public education. This responsibility requires the ability to manage all department programs at the company level. To accomplish their duties, captains must have the ability to interact efficiently with members of their organizations, department program managers, and the public.

Captains must have considerable knowledge of fire suppression and EMS skills. As company officers, they must be prepared to assume command of emergency incidents until relieved by a senior officer. Captains may be assigned to a number of staff functions, including training, communications, fire prevention, or resource management. While the duties of these positions will vary, all require strong leadership and communications skills as well as the ability to plan, schedule, and manage work.

Candidates for captain are advised to complete the following courses:

- Hazardous Materials
- Fire Investigation
- Fireground Tactics and Strategy
- Personnel Management for Firefighters
- Technical Writing
- Time Management
- Instructional Methods

These courses provide the basis for the technical knowledge that is required in a good officer. Additional education enhances an applicant's qualifications. An associates' degree is recommended for company officers, and members wishing to be promoted beyond the company officer level should have a baccalaureate degree.

Candidates for captain should prepare for supervisory positions early in their professional lives. Active participation as a firefighter, EMT, and apparatus driver/operator provides first-hand knowledge of the duties the captain supervises. Candidates for captain are better prepared for the position if they have served in staff assignments in more than one departmental area and possess a working knowledge of how these areas interact to accomplish the organization's goals.

Key Terms

Define in your own words the following terms:

certification _____

_____

codes _____

_____

professional _____

_____

standard _____

_____

QUESTIONS

Review Questions

1. List five of the roles of the company officer.

_____

_____

_____

_____

2. Where does the company officer fit into the fire department organization?

_____

_____

_____

_____

3. List some of the rewards of being a company officer.

_____

_____

_____

_____

4. List some of the challenges facing a company officer.

_____

_____

_____

5. What is the name of the document that identifies the competency standards for fire officers?

_____

_____

_____

_____

6. Why should fire officers be certified?

_____

_____

_____

_____

7. List the five strategies for success discussed in this chapter.

_____

_____

_____

_____

8. What is the company officer's primary job?

_____

_____

_____

_____

9. What is the company officer's role in leading the company?

_____

_____

_____

_____

10. What is the company officer's role in the department?

_____

_____

_____

_____

Discussion Questions

1. What knowledge do you expect a good company officer to possess?

_____

_____

_____

_____

2. What skills do you expect a good company officer to possess?

_____

_____

_____

_____

3. What abilities do you expect a good company officer to possess?

_____

_____

_____

_____

4. What behaviors do you expect a good company officer to demonstrate?

_____

_____

_____

_____

5. Are there any other qualities you expect to see in a good company officer?

_____

_____

_____

_____

6. Other than those listed in the text, what are the roles of a good company officer?

_____

_____

_____

_____

7. If you could design a course that addresses the items listed in questions 1–6, which would you include?

_____

_____

_____

_____

8. What are the qualities *you* personally need to develop to be a good company officer?

_____

_____

_____

_____

9. What are your personal (as opposed to the professional) goals? What are you doing to meet these goals?

_____

_____

_____

_____

10. What are your motives in becoming a company officer?

_____

_____

_____

_____

HOMEWORK FOR CHAPTER 1

The Promotion Process: Challenges and Opportunities for Officers

For Fire Officer I

Your assignment: Prepare a memo to your instructor providing your answers to the following questions:

1. What are the requirements for advancement to Fire Officer I in your fire department?
2. How do these requirements measure up with NFPA 1021?
3. How do you measure up with regard to these requirements as you start this course?
4. Reviewing the qualities of a good officer (Discussion question 3), what would you consider to be your greatest strength?
5. Reviewing the qualities of a good officer (Discussion question 3), what would you consider to be your greatest weakness?

For Fire Officer II

Your assignment: Prepare a memo to your instructor answering the following questions:

1. In your own words, how are officers promoted in your department?
2. Do you think the present system is adequate? Please explain your answer.
3. Regardless of the quality of the present promotion system, you must have some suggestions on ways to improve the system. What recommendations would you make to change the present process? Provide a detailed answer including specific suggestions that you would recommend.
4. If feasible, provide a copy of your department's SOP that affects promotion policy.

CHAPTER QUIZ

1. The company officer is:
   a. a first-line supervisor
   b. responsible for the safety of assigned personnel
   c. responsible for the performance of the company
   d. responsible for all of the above

2. Which of the following best describes the company officer?
   a. student
   b. manager
   c. teacher
   d. all of the above

3. In most departments, the typical company spends about _____ percent of their time dealing with emergency operations.
   a. 10
   b. 25
   c. 50
   d. 90

4. The company officer's primary job is to be a:
   a. coach
   b. instructor
   c. leader
   d. manager

5. NFPA 1021 lists the professional qualifications for:
   a. firefighter
   b. fire officer
   c. fire inspector
   d. fire investigator

6. The principal advantage of certification is that it:
   a. measures abilities against an established national standard
   b. protects the individual from liability
   c. provides justification for salary increases
   d. provides recognition of demonstrated proficiency

7. The professional standards identified in NFPA 1021 apply to:
   a. all officers in the fire service
   b. officers who have no other job
   c. paid fire officers
   d. volunteer fire officers

8. The company officer manages the department's resources:
   a. at the scene of emergencies
   b. at the fire station
   c. at fire department headquarters
   d. both a and b

9. In fire departments, approximately _____ percent of the employees work at the company level.
   a. 60–70
   b. 70–80
   c. 80–90
   d. 90–100

10. The fire officer plays an important role in providing a communications link for and to the:
    a. firefighters
    b. fire chief
    c. citizens
    d. all of the above

THE COMPANY OFFICER'S ROLE IN EFFECTIVE COMMUNICATIONS

 ## OUTLINE

OBJECTIVES

Upon completion of this chapter, you should be able to describe:
▎ The communications process
▎ The need for effective personal communications
▎ Ways to improve your oral communications
▎ Ways to improve your listening skills
▎ Ways to improve your writing skills

OVERVIEW

Good communications skills have a positive impact on every aspect of your work and personal life. Being able to communicate effectively enhances your leadership ability, helps you gain respect from your supervisors and peers, and makes you more effective in talking to the public, the media, and others. In all cases, your ability to communicate creates an impression about you, your organization, and your profession (see **Figure 2-1**).

Figure 2-1 Effective communications with employees, the organization, and the public are important parts of the officer's job.

Unfortunately, *poor* communications skills will also create an impression about you, your organization, and your profession. A lack of good communications skills can hurt anyone. For those in public service, poor communications skills hurt others as well. You must have effective communications skills to motivate your colleagues, influence your supervisor, and deal with the public you serve. Many officers overlook the need for developing effective communications skills. However, it may be one of the most important topics in the text. Work to be an effective communicator (see **Figure 2-2**).

COMMUNICATIONS ISSUES

A Needs Assessment of the U.S. Fire Service provides a wealth of information about the nation's fire service. Here are several items from that report pertaining to communications:

▌ Only 25 percent of the nation's fire department can communicate at an incident scene with their partners, the responders from other local and state emergency response agencies. The secretary of the Department of Homeland Security commented that this is *the major issue* for coordinating response efforts at the scene of large-scale operations.

▌ Less than 5 percent of the nation's fire departments have mobile data terminals.

▌ Only 2 percent of the nation's fire departments have advance personnel location equipment (GPS-based location systems).

▌ About 60 percent of the nation's fire departments reported having Internet access. Of these, 15 percent reported having access only at headquarters despite having multiple fire stations.

PRINCIPLES OF EFFECTIVE WRITING

Consider the reader. First and foremost, consider who your target reader will be. Use plain language. Use terms and abbreviations that your reader will clearly understand. If in doubt, spell out the term at its first use, followed by the abbreviation or acronym. Technical terms, jargon, and abbreviations are like hurdles on the track. You will find that they slow the reader rather than help.

Emphasis. Memos, letters, and directives should usually be limited to one topic. This practice allows the writer and the reader to focus on one issue and discuss it as needed until it is understood and resolved. This rule also helps satisfy the next item.

Figure 2-2 Good listening skills are part of communicating.

Brevity. Brevity is desired. Consider the material you receive in the mail every day. The letters that are brief and to the point usually get read right away. Brevity is often misunderstood. Brevity does not mean too short. It just means not too long, but long enough to do the job.

Simplicity. Use everyday words when possible, and use the same words you use while speaking. When writing, we frequently resort to bigger words, thinking they will impress the reader. Our efforts should impress the reader with the *information* we are sharing, not our writing style. In many cases, the use of larger words actually slows the communications process.

⬤ THE NEWS RELEASE

The news release is a standard form of communication used by organizations to alert the media to news pertaining to the organization's activities. News releases are often used to announce a new product or service. Public organizations, such as fire departments, often use a news release to announce events to which the public are invited.

The news release is usually printed on a form that makes it conspicuous (**Figure 2-3**). The text includes a description of the event; the date, time, and location of the first session; and any pertinent background information. To be successful, the news release should be no more than one page, and should be sent so that the information can be processed and published in time to benefit the public. Copies are normally mailed to the media (television, radio, and newspapers that serve your community).

A good news release will:

❚ Be on recognizable letterhead

❚ Contain the name, phone number, and e-mail address of a contact person

❚ Be simple and lack technical jargon

❚ Be well written and free of factual or grammatical errors

As with other forms of written communication, you should always proofread your work before making copies and mailing.

⬤ HOMEWORK ASSIGNMENTS

NFPA 1021, the *Standard for Fire Officer Professional Qualifications,* requires prospective officers to demonstrate their ability to effectively communicate. For those seeking certification, oral communications skills will be encouraged in the classroom portion of the course. Assessment of your written communications skills will be measured through written homework assignments.

DEPARTMENT OF INSURANCE
NEWS

Jim Long
Commissioner of Insurance

Department of Insurance
P.O. Box 26387
Raleigh, NC 27611

Media Advisory
May 5, 1995

Contacts: Paul Mahoney
Gena Arthur
(919) 733-5238

State Fire Marshal Promotes Safe Kids Week, May 6–13

RALEIGH—Unintentional injuries are the leading killer of children in the United States. Each year, more children ages 1 to 14 die from unintentional injuries than from all childhood diseases combined. Approximately 7,200 children ages 14 and under are killed and 50,000 permanently disabled each year. "We can work together to keep our children safe and reduce these numbers," said Jim Long, State Fire Marshal and Insurance Commissioner. "The *National* **SAFE KIDS** *Campaign* provides parents with important child safety information. Safe Kids Week may last just even days, but keeping our children safe is a year-round effort."

For children ages 1 to 4, fires and burns are the leading cause of unintentional injury-related death. "The chances of dying in a residential fire are cut in half when there is a working smoke detector in the home," said Long. "Ninety percent of child fire deaths occur in homes without working smoke detectors." You can keep your child from becoming a statistic. Install smoke detectors on every level of your home and near bedrooms; check your detector's batteries once a month and change the batteries twice a year. "The cost and maintenance of smoke detectors are a small price to pay for a child's life," said Long.

For children ages 5 to 14, motor vehicle injuries are the leading cause of unintentional injury-related death. "Seat belts save lives," said Long. "Forty-five lives were saved and 320 injuries prevented during the first six months of the "Click It or Ticket" campaign alone. Fewer deaths and injuries mean that taxpayers and insurance consumers will pay less to absorb the health and medical costs of high-risk drivers." Use a safety seat for children under age 4 and less than 40 pounds. For children 40 to 60 pounds use a booster seat. Teach older children to buckle up every time they get in a car.

Teach children to use bicycle helmets, also. Universal bike helmet use by kids ages 4 to 15 would prevent 135 to 155 deaths and 39,000 to 45,000 head injuries annually. "For children in this age group, every dollar spent on bicycle helmets saves $25 to $31 in medical costs and loss of future earnings and quality of life," said Long. "How much is your child's safety worth to you?"

For more information on Safe Kids Week and child safety, call the *National* **SAFE KIDS** *Campaign* at (202) 884-4993.

Figure 2-3 Sample news release.

MEMORANDUM

TO: **Fire Officer I Student**

FROM: **Clinton Smoke, author**

SUBJECT: **The Memorandum**

DATE: **February 7, 2004**

The business memorandum, or memo, is used to communicate with other people within the same organization. A memo can be quite long but usually it is brief. It is used to report on the progress of a project, raises or answers questions, or provides an opinion. Both the form and the structure of the memo make it quick and easy for both the preparer and the reader. As a result, it is widely used in business and government.

Since the memo is sent to another person in the organization, address, salutation or close are not needed. Instead, the addressee's and sender's name, and a subject are usually set out as shown above. Since the purpose of the memo is to provide a brief report, one should engage the reader at the very beginning. State the reason for writing in the first sentence and provide the answer, conclusion or evaluation in the first paragraph. The balance of the memo can be used to explain the basis for your conclusion. When closing, you should make a cordial offer to keep the communications process open if necessary.

While the memo can be less formal than a business letter, you should be careful to avoid technical terms and other jargon that might not be well understood. Most abbreviations take just about as long to read as the word or words they represent. Remember also that while you and the reader may well understand the subject, copies of memos are frequently passed to others to keep them informed, or filed for historical purposes. Try to write so that all of the readers will be able to understand your message.

I hope that this information is useful and that it will help you to appreciate the value of the memorandum as a form of written communication. You can find many good suggestions in books on technical or business writing, available in libraries and bookstores everywhere.

Please let me know if I can be of any additional assistance.

Figure 2-4 A sample memorandum.

What the NFPA 1021 Says

General Prerequisite Skills for Fire Officer I

"The ability to effectively communicate in writing utilizing technology provided by the AHJ (fire department); write reports, letters, and memos utilizing word processing and spreadsheet programs; (and) operate in an information management system."

Throughout the Standard we see such words as "The ability to communicate orally and in writing." In fact, the word *communicate* appears 15 times in the section of NFPA 1021 for Fire Officer I.

As you move up the leadership ladder to positions of increasing responsibility, you will be confronted with increasing requirements for effective writing. The lack of good writing skills tends to hold back good ideas and good people. As a result, individuals and organizations do not produce at their full potential. This problem is widespread, and there is considerable need for helping individuals work on their communications skills. This course provides opportunities for developing these skills.

Writing should be viewed as an opportunity rather than a burden. In the workplace, the task of writing is often made easier by the use of forms. Many of your written efforts involve the simple completion of a form by providing comments where appropriate. A bit more challenging is writing a memorandum or a letter (see **Figures 2-4** and **2-5**). To satisfy the requirements of NFPA 1021, you will be given an opportunity to write several memos and letters during this class.

NORTHERN VIRGINIA COMMUNITY COLLEGE

May 26, 2005

Lieutenant R. U. Ready
Plain City Fire Department
123 Main Street
Plain City, USA 34567

Dear Lieutenant Ready:

This letter is in response to your questions about the form and content of a business letter. A business letter is appropriate when writing to someone outside of your own organization. While there are many reasons for letters, they all have one thing in common: each is a direct communication between two people.

Letters are a valuable and lasting form of formal communications and can be an effective means of obtaining desired action. To be most effective, letters should get directly to the point, and then fill in the details as needed. The letter should close on a note of goodwill.

You should start by stating, clearly and directly, your purpose in writing. It may be appropriate to mention briefly any prior correspondence or why the recipient was selected. In a letter of application, for example, the first sentence might state that you are applying for a particular job. For the reader, who may be a staff person processing hundreds of letters for several vacant positions, they will know the purpose of your letter.

An additional paragraph or paragraphs may be needed to fill in the details or to justify, clarify, or emphasize your point. The information should be presented in a logical sequence that will help the reader draw the desired conclusion or take the desired action. The length of this section will depend upon the subject of the letter and the amount of information to be presented.

The final section of the letter is usually a short paragraph that indicates that the letter is about to close, and at the same time makes an effort to build goodwill.

Many reluctantly approach the writing of a business letter, yet we know that a good letter will often get us what we want. Writing an effective letter may benefit you personally or help your organization. So give it a try; I think you will be pleased with the result. You can find additional information on this topic in many training programs, college classes, and in books on writing available from libraries and bookstores. Good luck!

Sincerely,

Clinton Smoke

Clinton Smoke

Figure 2-5 Sample of a business letter.

In addition to meeting the requirements for certification, this process should help you improve your business communications skills. Your memo or letter will be read by the instructor and returned to you. Hopefully, your instructor will offer constructive suggestions regarding your writing skills.

Please realize that most of the instructors who teach fire science courses were not English majors in school, and they may not grade your work as completely as an English instructor would. However, they can read your work and answer two basic questions:

1. Did you complete the assignment?
2. Did you communicate effectively?

We suggest that you write as if you were writing for your supervisor. We also suggest that instructors temporarily place themselves in the role of your supervisor while reading your assignments.

This is not an English-writing course. However, you can meet the course objectives while improving your writing skills through practical and relatively painless writing exercises. Such a process provides an effective learning experience.

Several of the homework assignments will motivate you to think about the topic for the coming week. Hopefully, you will be encouraged to read the text and think about the topic before the class starts. Maybe you will be able to find a relevant article from one of the current professional journals. Bring the magazine to class and be prepared to briefly comment on the topic. You are an adult. Your teacher should not have to teach the class alone.

Other homework assignments will provide practical applications of things discussed in class. Most of the assignments are designed to satisfy a particular requirement in NFPA 1021.

Homework should be neatly typed in your organization's proper memo or letter form. If you are not sure of the exact style for preparing correspondence, you may use the samples on the following pages for guidance in submitting the homework.

Except where indicated otherwise, your memo or letter should be addressed to your instructor and written as if you were writing to your supervisor. However, to reinforce the comment in the text about avoiding jargon and abbreviations in your letters or memos, we suggest that you avoid abbreviations, local lingo, and terms not generally understood by the public.

The length of your memo or letter will be determined by the topic and your personal writing style. Again, the sample may provide some guidelines. You should be able to write 200 to 300 words (approximately a page) on any of the required topics. More than a half page is a minimum; continuing to a second page, if needed, is always acceptable.

 # REVIEW YOUR WORK

Writing for class should be approached in the same way you would write at work. Your answers are important, of course, as is the way you express yourself. Take time to carefully prepare your reply. Use these homework activities as an opportunity to improve your communications skills while stating your goals for this course.

Remember to avoid waiting until the last minute to do your work. The pressure of a deadline will force you to make compromises that will not be in your best interest. By doing your writing in advance, you also have the opportunity to review and edit your work before submitting it.

With computers in widespread use, the ability to use word processing software is available to nearly everyone. Use this technology to enhance your writing. Start by writing your letter or memo using the techniques we have suggested. Review it on the screen and run the spell check. Then print a draft copy, read it, and mark up the copy with corrections on areas where you think it needs improvement. Make those corrections and print another copy.

Wait a few hours or until the next day if possible and repeat the process. In nearly every case, you will make additional corrections when you reread the document. You will add a word or two or rearrange several sentences to improve your work. Remember, you are trying to enhance the reader's understanding. Make the corrections and print another copy. If time permits, repeat the process.

Although your second or third reading is always beneficial, it is also beneficial to ask someone else to read the document. That person does not have to be an editor or a grammar teacher, but someone who has good communications skills and who has the patience to read very carefully. Good proofreading is not like reading the sports pages of the paper; it requires slow, deliberate reading.

Some writers read aloud when proofreading. In doing so, they hear mistakes more readily than reading in the usual manner. In any case, you or someone else should read the document carefully and make corrections where necessary. Good writers perform both for every document. These steps will significantly enhance your writing.

Key Terms

Define in your own words the following terms:

active listening_____

_____

barrier _____

_____

feedback _____

_____

medium _____

_____

message _____

_____

receiver _____

_____

sender _____

_____

QUESTIONS

Review Questions

1. What is the communications process?

_____

_____

_____

_____

2. What is the role of the company officer in communications?

_____

_____

_____

_____

3. What are the five parts of the communications model?

_____

_____

_____

_____

4. What are the common barriers to communications?

_____

_____

_____

_____

5. How can these barriers be minimized?

_____

_____

_____

_____

6. What is meant by the term _active listening_?

_____

_____

_____

_____

7. How does your communications style indicate your leadership style?

_____

_____

_____

_____

8. Name several of the principles of effective writing.

_____

_____

_____

_____

9. What steps can you take to improve your writing?

_____

_____

_____

_____

10. What is the company officer's role in the communications process?

_____

_____

_____

_____

Discussion Questions

1. What are the benefits of effective communications?
   a. At work

_____

_____

_____

_____

   b. At home

_____

_____

_____

_____

2. When writing, what steps can you take to enhance the finished product?

_____

_____

_____

_____

3. How do improved communications skills help you in your professional life?

_____

_____

_____

_____

4. How do improved communications skills help you in your personal life?

_____

_____

_____

_____

5. List any barriers to effective communications found in your organization or workplace.

_____

_____

_____

_____

6. How has the communications process changed the way emergency response organizations are notified of an emergency?

_____

_____

_____

_____

7. How has the communications process changed for emergency response organizations in the way they are dispatched to the emergency?

_____

_____

_____

_____

8. How has the communications process changed the way emergency response organizations communicate at an emergency scene?

_____

_____

_____

_____

9. While communications technology has improved in recent years, many issues are still apparent. What are the issues in emergency communications in your community?

_____

_____

_____

_____

10. What do you think the future will bring in the way of changes in communications technology?

_____

_____

_____

_____

HOMEWORK FOR CHAPTER 2

For Fire Officer I

A Citizen's Complaint

What NFPA 1021 says: Section 4.3.2. The Fire Officer I shall "initiate action to a citizen's concern, given policies and procedures, so that the concern is answered or referred to the correct individual for action and all policies and procedures are complied with."

Situation:
A citizen recently complained to one of the community's elected officials about noise from your station. Sirens and air horns were mentioned but the real criticism seems to be the nearly continuous noise from the external speakers at your fire station. She reports hearing personal paging and dispatch information from the speakers 24 hours a day, even when no one is at the station. The citizen's comment was passed to the chief. The chief called you (company officer) and asked you to look into the problem and "bring me a solution."

Your assignment:

1. Investigate what can be done to resolve the problem. For purposes of this exercise, you can use your imagination, but be realistic.
2. Prepare a memo to the chief outlining your findings and your recommendations for correcting the problem.
3. Draft a business letter for the fire chief's signature. Address the letter to: Ms. Dorothy Jones, 201 Firehouse Court, Anytown, Virginia, 23456. The letter should respond to the citizen's complaint.

To summarize, this assignment asks you to prepare two documents. One is a brief memo report to the chief. Attached to the memo should be a polite letter to your neighbor for the chief to sign. Submit both documents to your instructor.

For Fire Officer II

A News Release

What NFPA 1021 says: Section 5.4.4. The Fire Officer II shall "prepare a news release, given an event or topic, so that the information is accurate and formatted correctly."

Your assignment: Prepare and submit to your instructor a news release about this course and about your participation in this course. Information and a sample are in this manual.

CHAPTER QUIZ

1. Communications can be defined as the:
   a. exchange of ideas or information
   b. exchange of words
   c. information given over the radio
   d. writing or talking to another
2. In communications, the information being transmitted is known as the:
   a. medium
   b. message
   c. receiver
   d. sender

3. The proper sequence for the steps in the communications model are:
   a. sender, message, receiver, feedback
   b. message, receiver, feedback, sender
   c. receiver, feedback, sender, message
   d. feedback, sender, message, receiver

4. Walls, distance, and background noise are examples of things that:
   a. provide for effective communications
   b. act as physical barriers to effective communications
   c. act as personal barriers to effective communications
   d. act as semantic barriers to effective communications

5. The deliberate process of focusing one's attention on the communications of another is called:
   a. active listening
   b. listening
   c. observation
   d. passive listening

6. Which of the following is the most common form of communications?
   a. formal correspondence
   b. memos and e-mails
   c. spoken communications
   d. written communications

7. Feedback is the:
   a. barrier to communication
   b. information being sent
   c. verification that the message was received and understood
   d. way a message was sent

8. Any obstacle in the communication process is referred to as a:
   a. barrier
   b. block
   c. problem
   d. wall

9. Effective communication has taken place when:
   a. there is appropriate feedback
   b. the sender uses the proper medium
   c. the message is received
   d. the message is sent

10. Which of the following is an example of nonverbal communication?
   a. frown
   b. policy manual
   c. posted safety procedure
   d. stop sign

THE COMPANY OFFICER'S ROLE IN THE ORGANIZATION

 OUTLINE

- Objectives
- Overview
- Communicating Policy in Organizations

- Key Terms
- Review Questions
- Discussion Questions

- Homework
- Chapter Quiz

 OBJECTIVES

Upon completion of this chapter, you should be able to describe:

- Organizational structure
- Authority and responsibility
- Lines of authority
- Duties and responsibilities
- Line and staff organizations
- The role of the company officer in the organization

 OVERVIEW

The company officer has many roles in the organization. First, as a company officer, you represent the fire department to the average citizen. The company officer is often the first on the scene, the last to leave, and for many events is the only person with whom the customer talks.

At larger events, other units of the department may be represented. Other companies may be present, and other parts of the organization may support the operation, whether or not they are at the scene of the emergency. Company officers must understand how the entire organization works, how his or her company fits into the total organization, what resources are available and appropriate for any type of incident, and how to help customers connect with the appropriate fire service organization to receive needed services.

As a company officer, you are a vital part of the organization that you represent. You should understand your role, how you represent your company and its needs to the department's management team, and how you represent the rest of the organization to the firefighters within your company (see **Figure 3-1**).

Figure 3-1 For members of the team, the officer is a teammate and a link to the rest of the organization.

Understanding organizations is important. In the past, people worshiped organizational structure, paid homage to the chain of command, and measured organizations by the number of rules they had. Those are good things, and we should understand their value, both in today's organizations and for those who grew up in this culture. However, we are learning that there may be better ways to run an organization (see **Figure 3-2**). We are seeing more organizations move away from rules and structure and focus instead on values, especially where the employees are concerned.

COMMUNICATING POLICY IN ORGANIZATIONS

If you called headquarters and requested guidelines for every aspect of the job, you would probably receive slightly different advice each time you called. It is likely that others, asking similar questions, would receive slightly different answers as well. With a written document, everyone can be provided with the same answer, and we can usually find the answer without making a phone call. For most items, once you know the policy, you do not have to look it up again.

Most well-managed organizations provide information to their members in a variety of ways. This information helps improve efficiency and keeps everyone headed in the same general direction. Some of this information has a long life. These are called *standing documents*.

Guideline policies and procedures[1] are examples of standing documents. Standing documents provide guidelines for events that occur frequently. Rather than establishing policy on an event-by-event basis, a standing document provides direction on how to proceed. For fire departments, these documents cover everything from actions for first-arriving companies at the scene of a major fire to ordering housekeeping supplies for the fire station.

Standing documents provide a means to publish *policy*. Policy may start at the top of an organization. A policymaker (manager) may want to provide and publish available information for those in the field. The policy will provide direction for action and usually sets boundaries for that action. Situations that do not fall within the boundaries are referred up the organizational chain of command to the appropriate level for resolution. Regardless of the means or quantity of material that is provided, such documentation does have its advantages.

---

[1]The terms guidelines, policies, and procedures, are terms that are used somewhat interchangeably. In many departments, the guidelines are the broadest policy statements, followed by policies and procedures. Procedures tend to be very limited in scope and very specific. For example, a procedure might address how to load hose on a fire apparatus.

Figure 3-2 Effective management of any organization involves the ability to delegate certain tasks. In many cases, these tasks are delegated to company officers.

Other policy documents may be the result of an event or a question that arises from within the organization. In either case, the question is referred up the chain of command to someone with the appropriate authority who can make a decision. Some organizations will answer the question and stop there. Others, being a little more proactive, will realize that similar events are likely to occur in the future and will take the initiative and publish a policy statement on how to react when the situation occurs again.

Obviously, we do not need to publish a directive every time we make a decision. However, we do not want to make the same decisions over and over. Somewhere in between is a reasonable middle ground. Finding that middle ground is the mark of good management.

Some organizations pride themselves in the amount of documentation they provide. Others pride themselves on how little they have. Some are getting away from a paper environment and provide their directives "online" through networked computers.

Where this method of information management is not available, many departments have converted their paper documents to disks or other files that can be viewed on the computer. Disks take less storage space than paper records, and finding a particular bit of information is often easier and faster.

Management's intent is usually clearer with written policy. Well-written policy will be reviewed by most of the department's senior staff. Problems, conflicts, and ambiguities within the document should be cleared up before the document is published.

Advantages and Disadvantages of Written Policy

There are benefits to providing good, clear policy guidelines:

■ They allow the same information to be transmitted to several locations and assure that all work sites have the same policy guidance.

■ They provide a written record that will be available in the future.

■ They provide evidence that the policy was established and that everyone was informed.

■ They allow the information to be fully understood before the receiver reacts.

Clearly, there are some disadvantages of having the answers on paper:

▌ The information takes longer to publish.

▌ Writing takes more skill than speaking.

▌ Written communications do not provide immediate feedback.

▌ Written communications do not allow for feedback such as lack of facial expressions, gestures, and so on.

▌ Written communications tend to be longer and more complicated than verbal answers.

Types of Policy Documents

Orders

Orders give specific direction. They assume that the member understands a situation and has the training and equipment to deal with it.

Directives

Directives are not as specific. They are more like a goal and leave to the company officer or other supervisor the task of getting the job accomplished. For example, a directive may state that the fire station must be kept clean and orderly. This provides the company commander with a great deal of leeway in defining the acceptable standards of cleanliness and orderliness and how the directive should be followed.

We frequently complain about the directives we receive. We should understand that the process of writing a clear, effective directive is not always an easy task. Those in staff positions should always invite comments and questions.

Generally, standing documents do not list names of individuals since these change. Rather, they identify the position. The incumbent should understand what this implies. An exception is to place a name somewhere in the document where one can get more information. Typically, this is the last line, something like "For more information, please call Ms. A. B. Seay in the Human Resources Section at 234-5678."

General topics for standing documents are (not necessarily in order of importance):

▌ Communications Procedures

▌ Standard Procedures for Emergency Operations

▌ Headquarters Functions

▌ Organizational Structure and Relationships

▌ Personnel Administration

▌ Prevention Programs

▌ Regulations for Personal Conduct

▌ Station Management

▌ Use of Facilities and Equipment

These should be sorted into some sort of filing system so that the information can be easily found.

Distribution and Storage

Usually, the administrative procedures of the department cover the publications and distribution process for these documents. A good department will distribute this information in a routine manner so that personnel, especially at more remote sites like fire stations, can be assured that they are getting all documents in a timely manner.

Directives should be stored in an appropriate place, usually in a three-ring binder. However, some routine method of notifying personnel that a new directive has been received should be established. Many organizations are moving to a paperless environment in which all policy documents and forms are now promulgated and stored electronically. This approach has many advantages: it provides a quick, complete and automatic distribution of all policy documents, it saves a lot of paper, and it provides all users with the latest version of information.

Examples of Standing Documents

Standard operating procedures. A standard operating procedure (SOP) establishes a process that will be followed by all members. They are usually part of a continuing program and remain valid until changed or cancelled. Deviations are not permitted except under the most unusual circumstances.

General orders. A general order is a directive that tells one or more people to take some specific action. General orders cover a specific period of time which is usually spelled out in the directive.

Operating manuals or instructions. Operating manuals or operating instructions are comprehensive documents that cover complex subjects. They are like SOPs but, because of their size, usually exist as stand-alone documents. For example, a large department might have a manual dedicated to the organization's incident management system.

All of the foregoing standing documents require compliance. Some of these documents affect everyone, while some may affect only a few individuals.

Other documents may be issued. Training bulletins, safety bulletins, and general informational bulletins are commonly used. These are usually one- or two-page documents intended for a wide distribution. They cover one topic and are used for information only. Unlike standing documents, they are not permanent and are automatically cancelled.

Writing Policy Documents

All of the documents discussed here are examples of formal communications. All should be examples of good clear writing. The real test of quality writing comes when the paper hits the street. If personnel at the stations can understand and accomplish what a directive says, it is a good indication that the document was well written. If, however, we receive many phone calls from confused troops, the writer might be called into the chief's office and the document must be rewritten. It would appear that we did not do very well.

In addition to previously listed techniques for good writing, a document of this importance should be circulated to a test audience of affected users. For example, if the proposed document provides some policy for firefighting in a particular occupancy, circulate the draft document to a representative group of company officers for their comments. To expedite this process, send it to the entire group concurrently, and give them a reasonable but specific deadline for their comments to be returned.

It is usually necessary to review policy documents from time to time to see if they are still valid. Some organizations have a built-in cancellation process that kills the document at a specific age, for example, five years. All organizations should review their policies on a regular basis and remove those that are no longer needed.

It is also a good idea to regularly publish a list of the in-effect instructions so station officers and other supervisors can verify that their directives are current and complete.

Company officers can also use this information at the company level. If a policy needs to be established for station activities, organize the information in a format that is consistent with other department communications. Give others a chance to review it and ask questions before it becomes "the law"!

Key Terms

Define in your own words the following terms:

accountability _____

_____

delegate _____

_____

delegation _____
_____

discipline _____
_____

line functions _____
_____

line authority _____
_____

public fire department _____
_____

responsibility _____
_____

scaler principle _____
_____

span of control _____
_____

staff functions _____
_____

staffing _____
_____

unity of command _____
_____

QUESTIONS
Review Questions

1. Define *organization*.

_____
_____
_____
_____

2. Why do we need structure in organizations?

_____

_____

_____

_____

3. How do organizational charts show lines of authority?

_____

_____

_____

_____

4. How do organizational charts show lines of communication?

_____

_____

_____

_____

5. Define *division of labor.*

_____

_____

_____

_____

6. Define *span of control.*

_____

_____

_____

_____

7. Define *unity of command.*

_____

_____

_____

_____

8. What is meant by *delegation*?

_____

_____

_____

_____

9. Distinguish between "line" and "staff" in an organization.

_____

_____

_____

_____

10. What is the role of the company officer in the organization?

_____

_____

_____

_____

Discussion Questions

1. What type of local government do you have where you live/work?

_____

_____

_____

_____

2. Describe the composition of the fire service in the United States.

_____

_____

_____

_____

3. What are the challenges facing the volunteer fire service today?

_____

_____

_____

_____

4. The statistics in figures 3-4 and 3-5 on page 58 of the textbook suggest that total call volume is up while the number of fires is down. What is making up the difference? Will this trend continue?

_____

_____

_____

_____

5. The statistics referred to in question 4 represent the national picture. What is happening in your community?

_____

_____

_____

_____

6. What is the legal authority for fire departments where you live?

_____

_____

_____

_____

7. What are the roles of your department's senior officers?

_____

_____

_____

_____

8. Where are these roles defined?

_____

_____

_____

_____

9. What are the roles of the company officer within the department?

_____

_____

_____

_____

10. What is the company officer important to the organization?

_____

_____

_____

_____

HOMEWORK FOR CHAPTER 3
The Fire Department's Organization

For Fire Officer I

What NFPA 1021 says: 4.1.1. The Fire Officer I shall "explain the organizational structure of the department; geographical configuration and characteristics of response districts; departmental operating procedures for administration, emergency operations, incident management systems, and safety; . . . and policies and procedures regarding the operation of the department as they involve supervisors and members." NFPA 1021 goes on to explain that for the purposes of this standard, the term *fire department* shall include any public, private, or military organization engaging in this type of activity.

Your assignment: Obtain a copy of your department's organization chart. Prepare a memo to your instructor addressing in complete sentences each of the following questions:

1. How accessible was the organization chart to you?
2. Does the chart appear to be up-to-date?
3. Does the chart clearly define the lines of authority?
4. How many organizational layers are there between you and the fire chief?
5. If you were asked to create a new chart, what changes would you make?

Attach a copy of the organization chart to your memo.

For Fire Officer II

What NFPA 1021 says: 5.1.1 The Fire Officer II shall explain "the organization of local government; enabling and regulatory legislation and the law-making process at the local, state/provincial, and federal levels; and the functions of other bureaus, divisions, agencies, and organizations and their roles and responsibilities that relate to the fire service." NFPA 1021 goes on to explain that for the purposes of this standard, the term *organization* includes groups providing rescue, fire suppression, and other related activities.

Your assignment: Find the local authority for your department. If within a city, town, county, or fire district, your department is most likely authorized, either directly or indirectly, by state statute. Find that statute and provide a copy of the applicable section.

Also find a government organization chart that includes your fire department.

In a memo to your instructor, explain in your own words your fire department's authority for providing fire and rescue services. Attach a copy of the chart to your memo.

CHAPTER QUIZ

1. There are about _____ fire departments in the United States.
   a. 20,000
   b. 30,000
   c. 40,000
   d. 50,000
2. Activities that support the companies providing emergency service to the citizens are called:
   a. assignment
   b. staff functions
   c. secretaries
   d. support staff
3. Which of the following terms describes the number of persons one can effectively manage?

   a. delegation of labor

   b. division of labor

   c. hierarchy of management

   d. span of control

4. Which of the following describes the process of breaking large tasks into smaller ones?

   a. delegation of labor

   b. division of labor

   c. hierarchy of management

   d. span of control

5. The organizational principle whereby each employee has but one supervisor is called:

   a. chain of command

   b. division of labor

   c. scaler principle

   d. unity of command

6. The organization chart describes:

   a. lines of authority and responsibility

   b. the lines of communication

   c. the activities of various divisions of an organization

   d. all of the above

7. As used in this chapter, *discipline* is described as:

   a. a system of rules and regulations

   b. the authority of appointed leaders

   c. the process by which employees are punished

   d. none of the above

8. A civilian employed by a fire department to perform nonemergency duties such as fire and life-safety education represents a _____ position.

   a. line

   b. scaler

   c. staff

   d. subordinate

9. The flat organizational structure is most effective for the company when the tasks being performed are _____ tasks.

   a. simple

   b. interdependent

   c. not interdependent

   d. nonemergency

10. The organizational structure which is most effective for the company when problem solving is:

   a. circular

   b. flat

   c. scaler

   d. social

THE COMPANY OFFICER'S ROLE IN MANAGEMENT

 OUTLINE

- Objectives
- Overview
- Functions of Management
- Fayol's Principles
- Theory X and Theory Y

- Code of Ethics
- Guidelines for Ethical Decisions
- Ethics Self-Quiz
- Key Terms

- Review Questions
- Discussion Questions
- Homework
- Chapter Quiz

OBJECTIVES

Upon completion of this chapter, you should be able to describe:

- Management
- The functions of management
- Recent major contributions to management science
- The role of ethics in management

 OVERVIEW

The company officer is a supervisor, a leader, and a manager. This chapter, and the one following, focus on the company officer's leadership role. Management is the act of guiding an organization's human and physical resources to attain the organization's objectives. Management includes the determination of what needs to be done and the accomplishment of the task itself. Managers plan and make decisions about how others will use the organization's resources. They should strive to do this efficiently and effectively (see **Figure 4-1**).

As a company officer, you are part of the department's management team. In keeping with contemporary management trends, many departments are passing much of the management process down to the company officer and beyond. When permitted to do so, company officers can have a great impact on the effectiveness and efficiency of the entire organization.

Company officers should be aware of and use all functions of management. Modern officers should be aware of management style models that have evolved through management science and use them as needed.

As with other tools you use on the job, no one uses all of the tools all of the time. However, it is nice to know which tool is right for the job at hand and to know how to use it effectively.

Figure 4-1 For company officers, the greatest resource is your teammates. Effective coordination of their efforts is essential for company excellence.

Ethics are important for the company officer too. The best way to have good, fair, and honest people in the station environment is for the company officer to set a good personal example and to expect the same level of performance and behavior from all others.

FUNCTIONS OF MANAGEMENT

The text states that there are five functions of management. Various texts list four, five, and even six functions of management. When four are stated, we usually find them listed as planning, organizing, actuating, and controlling. When six activities are stated, they are usually listed as planning, organizing, staffing, directing, controlling, and evaluating. How many, really, are there?

In each of the lists, we find that two things are consistent: the start and the end. All discussions on management functions start with planning and end with a process by which activities are monitored and necessary corrections are implemented.

Differences in the lists are caused by terms that describe the intermediate steps. In most cases, the mental process is the same. Other times, actual management activity, organizing, and commanding are pretty much the same. Once the planning activity is complete, the process involves identifying resources, bringing the resources together, and, in the case of human resources, giving them direction.

Once this is accomplished, some means should be established for evaluating the process to determine if the original goals and objectives were met. Whenever a new project is undertaken, the impact of the change should be evaluated since there are likely to be positive and negative results. Often, it is difficult to measure these results. Well-stated goals help. Corrective action may be needed to reach those goals. In the list of five functions used in your text, this evaluation and the associated correction activity is called *controlling*.

FAYOL'S PRINCIPLES

1. **Division of labor.** Divide the work into manageable portions and specialize the activities to improve efficiency. In the Bible's earliest reference to management, this is what Jethro told Moses to do in Chapter 18 of the Book of Exodus.

2. **Authority and responsibility.** These terms go hand in hand. Where authority arises, responsibility must follow. Managers must have the authority to make decisions and give directions.

3. **Discipline,** as a characteristic of authoritarian organizations, includes good leadership and clear agreement between management and labor on the role of each.

4. **Unity of command** suggests that one should receive orders from only one supervisor.

5. A **unity of direction** suggests a commitment from all to focus on the same objectives.

6. A **subordination of individual interests** to general interests in an organization suggests that the general interest must take precedence over the individual interests of the members.

7. **Proper remuneration** means a fair wage for fair work.

8. A **centralization of authority,** like the division of work, brings order to the organization. It provides accountability and responsibility where necessary. However, decentralization (we might think of it more as delegating) usually increases the worker's importance and value.

9. A **scaler (continuous) chain of ranks,** or layers in an organization, is the notion that authority flows from the highest to the lowest ranks in the organization's chain of command.

10. **Order** means a place for everything and everything in its place. Fayol was interested in material order and social order. Today, we say that order means having the right personnel and the right materials ready to do the job so that members are effective and efficient.

11. **Equity and fairness** are important in the treatment of the members of any organization.

12. **Initiative** is important to the organization and its members. The ability to think out a plan and see it through to completion is one of man's greatest satisfactions.

13. **Stability of our personnel** means recognizing that there is value in a long-term relationship with our members.

14. **Esprit de corps,** an organizational spirit, is essential to the survival of any organization.

THEORY X AND THEORY Y

A manager's style is influenced by concern for people in general and for members in particular. McGregor described this behavior and provided labels for the extreme ends of the spectrum of management style. He used the term *Theory X* to describe managers who assume that the average member is indolent, lacks ambition, dislikes responsibility, and prefers to be led. According to McGregor, Theory X-style managers feel that subordinates must be closely supervised.

In contrast, *Theory Y* managers see the potential in people, their capacity for assuming responsibility, and their readiness to strive to achieve organizational goals. According to McGregor, Theory Y managers will provide an environment where members can recognize and develop these characteristic for themselves (see **Figure 4-2).**

CODE OF ETHICS

1. Be loyal to the highest moral principles, and place allegiance to country above loyalty to any person, party, or government department.

2. Uphold the Constitution, laws, and regulations of the United States and all of the governments therein.

3. Give a full day's labor for a full day's pay, giving earnest effort and best thought to the performance of duties.

4. Seek to find and employ more efficient and economical ways to accomplish tasks.

5. Never discriminate by dispensing special favors or privileges to anyone, and never accept favors or benefits under circumstances that might be construed by reasonable people as influencing the performance of duties.

6. Make no private promises of any kind that are binding on the duties of your office because your private word cannot be considered as binding on your public duty.

7. Engage in no business that is inconsistent with the conscientious performance of duties.

Deming's Fourteen Points (Altered a Bit for the Fire Service)

1. Understand and accept the organization's mission, goals, and objectives.
2. Seek ways to improve the process and train people to do it better. All improvements, big and small, are important.
3. Use inspections as an improvement tool, not a threat.
4. Think beyond the bottom line. Consider quality and personal value along with the cost in every decision.
5. Demand quality, not better numbers, in everything you do.
6. Accept the idea of quality. Do every job right, the first time, every time.
7. Use training time effectively; some organizations train just to meet quotas.
8. Promote pride in the job. As supervisors, be sure that your members have the equipment and materials they need and recognize their good work.
9. Promote leadership. Work on giving supervisors the best tools possible to deal with the organization's most valuable resource—its people.
10. Create an atmosphere of trust and open communication.
11. Get rid of the organizational barriers that inhibit trust and effective communications, and force people to cooperate rather than compete.
12. Use analytical tools to better understand and monitor the job.
13. Work as a team.
14. Use TQM all the time.

Figure 4-2 Good company officers get their teammates involved with management functions. It makes for a better, more productive workplace and helps teammates grow professionally by learning new skills.

8. Never use any information gained confidentially in the performance of duties as a means to make a private profit.

9. Maintain high standards of personal integrity, honesty, and straightforwardness.

10. Provide fair and equal lifesaving service to all.

11. Expose corruption wherever you discover it.

12. Uphold these principles, ever conscious that a public position is a public trust.[1]

GUIDELINES FOR ETHICAL DECISIONS

1. Is it legal?
2. Does the act or behavior make sense?
3. Will people in authority approve of your decision?
4. How will others react to your decision?
5. Would you want your family to know what you have done?
6. Would you want to read about your decision in the papers?

ETHICS SELF-QUIZ

Please honestly answer the following questions.

Profession-Related Questions

Yes	No	
☐	☐	1. Do you share recognition with others when praise is given to you by your supervisor?
☐	☐	2. Are you honest with your fire department with regard to the use of the organization's supplies, telephone, fax machine, and copy machine?
☐	☐	3. Are you honest with your fire department in the accurate reporting of your working hours?
☐	☐	4. Are you honest with your fire department in the accurate reporting of travel claims?
☐	☐	5. Are you fair to your subordinates by making sure they receive complete and timely information that can help them in their careers?
☐	☐	6. Are you fair to your subordinates in reporting their work accomplishments in their performance appraisal process?

Personal Questions

Yes	No	
☐	☐	7. Are you honest with yourself and your spouse?
☐	☐	8. Do you submit complete and accurate information for your income tax report?
☐	☐	9. Do you obey the laws while driving your own vehicle?
☐	☐	10. Do you accept compensation from more than one source for the same work?
☐	☐	11. Do you conduct personal business activities while working for the fire department?
☐	☐	12. Have you been honest with yourself in answering the questions on this test?

---

[1]From "A Code of Honor" by Robert D. Carnahan, *Fire-Rescue Magazine,* January 1998, p. 78. Reprinted with permission.

Ethics Situations

In each of the following situations, you are the company officer. How would you handle each situation? Several choices are offered. You can pick one of these or take another course of action. Explain your choice.

A. Shortly after coming on duty, your company is dispatched to a working fire. You are the officer on the second arriving engine company. You find the officer of the first arriving company sitting on the curb, displaying signs of a hangover. He is a 25-year veteran and was your first company officer.

Do you:

1. Tell him to straighten out and get back to work?

_____

_____

_____

_____

2. Tell him to go get in the ambulance?

_____

_____

_____

_____

3. Ignore him and work with other personnel to attack the fire?

_____

_____

_____

_____

4. Call the en route battalion chief, describe the situation, and ask for instructions?

_____

_____

_____

_____

5. Take other action?

_____

_____

_____

_____

B. In responding to a motor vehicle accident, you discover that one of the victims is the son of a female firefighter who is on duty at your station. The child is critically injured.

Do you:

1. Try to get in contact with the father?

_____

_____

_____

_____

2. Call the station on the radio and explain what happened?

_____

_____

_____

_____

3. Send someone to bring the mother to the scene?

_____

_____

_____

_____

4. Personally go to the mother and explain what happened?

_____

_____

_____

_____

5. Take other action?

_____

_____

_____

_____

C. The wife of one of your firefighters calls you while you are on duty. She tells you that her husband is a heavy drinker and beats her while he is drinking. She wants you to warn him that if he does it again, she will go to the police.

Do you:

1. Advise the wife that this is a private matter and that you cannot get involved?

_____

_____

_____

_____

2. Offer the wife information about where she can get help and counseling?

_____

_____

_____

_____

3. Talk to the firefighter and tell him of the conversation?

_____

_____

_____

_____

4. Advise all of your firefighters of the department's member assistance programs?

_____

_____

_____

_____

5. Take other action?

_____

_____

_____

_____

D. You are the new company officer in an area where the owner of a local ice cream parlor frequently gives an extra dip of ice cream to on-duty firefighters when they stop by. The department's policy on accepting gifts is quite clear: Do not accept money, gifts, or presents while acting as a representative of the department.

　　Several members of your company take full advantage of this perk. On a warm day, while returning from a call, your driver suggests the company stop and "fill up on ice cream."

Do you:

1. Allow the visit?

_____

_____

_____

_____

2. Allow the visit but refuse to participate?

_____

_____

_____

_____

3. Allow the visit but talk to the owner about the practice?

_____

_____

_____

_____

4. Ask the driver to continue to the station?

_____

_____

_____

_____

5. Take other action?

_____

_____

_____

_____

E. As you are about to leave the station at the end of a shift, you observe one of your firefighters removing a tool box from another firefighter's pickup truck and placing it in the trunk of his own car. When you return to the station several days later, the firefighter who owns the pickup complains about the missing tool box, saying that he thinks someone at the station stole it.

Do you:

1. Assume it is a practical joke and ignore it?

_____

_____

_____

_____

2. Speak up and tell the owner what you know?

_____

_____

_____

_____

3. Say nothing but later talk to the firefighter who took the tool box?

_____

_____

_____

_____

4. Get your crew together and talk about the situation?

_____

_____

_____

_____

5. Take other action?

_____

_____

_____

_____

F. While having dinner with your family at a nice restaurant, you make a trip to the restroom and discover that one of the exit doors is chained shut.

Do you:

1. Do nothing?

_____

_____

_____

_____

2. Leave without doing anything?

_____

_____

_____

_____

3. Ask the manager to come to your table and point out the problem?

_____

_____

_____

_____

4. Call the fire marshal on duty?

_____

_____

_____

_____

5. Take other action?

_____
_____
_____
_____

Key Terms

Define in your own words the following terms:

commanding _____
_____

controlling _____
_____

coordinating _____
_____

customer service _____
_____

ethics _____
_____

Fayol's bridge _____
_____

internal customers _____
_____

management _____
_____

MBE _____
_____

MBO _____
_____

organizing _____
_____

planning _____

_____

policy _____

_____

procedure _____

_____

Theory X_____

_____

Theory Y_____

_____

Theory Z_____

_____

TQM _____

_____

QUESTIONS
Review Questions

1. Define *management*.

_____

_____

_____

_____

2. What are the functions of management?

_____

_____

_____

_____

3. Among the functions of management, planning seems to get a lot of attention. Why is planning so important?

_____

_____

_____

_____

4. Are company officers managers?

_____

_____

_____

_____

5. What resources are available for management by company officers?

_____

_____

_____

_____

6. Describe the characteristics associated with Theory X, Y, and Z.

_____

_____

_____

_____

7. Who are the customers of a fire department?

_____

_____

_____

_____

8. What resources are available to assess the quality of a fire department?

_____

_____

_____

_____

9. What is meant by the term _ethics_?

_____

_____

_____

_____

10. How does ethics tie in with the company officer's position?

_____

_____

_____

_____

Discussion Questions

1. How does management theory apply to the fire service?

   _____

   _____

   _____

   _____

2. Discuss Fayol's principles in the context of a modern fire department.

   _____

   _____

   _____

   _____

3. Which of Fayol's functions would be represented in each of the following?
   Selecting the site for a new fire station

   _____

   _____

   _____

   _____

   Arranging resources at the scene of an emergency event

   _____

   _____

   _____

   Examining the most efficient use of existing apparatus

   _____

   _____

   _____

   _____

   Determining the cause of a gradual increase in response time

   _____

   _____

   _____

   _____

Assigning tasks to accomplish the daily needs of a fire station

_____

_____

_____

_____

4. What lessons can be learned from a study of management theory?

_____

_____

_____

_____

5. How have management practices changed over the past 50 years?

_____

_____

_____

_____

6. How do you think management practices will change over the next 10 years?

_____

_____

_____

_____

7. The text discussed several ways to assess a fire department's capabilities. What value is there in using such external benchmarks in measuring such an organization?

_____

_____

_____

_____

8. How do you feel about ethics in public service organizations?

_____

_____

_____

_____

9. Why are ethics important in such organizations?

_____

_____

_____

_____

10. Explain the management roles of a company officer.

_____

_____

_____

_____

HOMEWORK FOR CHAPTER 4

For Fire Officer I

Respond to a Citizen's Concern

What NFPA 1021 says: Section 4.3. "This duty involves dealing with inquiries of the community and projecting the role of the department to the public and delivering safety, injury, and fire prevention education programs, according to the following job performance requirements." Specifically, section 4.3.4 says that the Fire Officer I shall "respond to a public inquiry, given policies and procedures, so that the inquiry is answered accurately, courteously, and in accordance with applicable policies and procedures."

Situation:
A citizen has asked the fire chief for information about the department's activities. The chief has asked you, in turn, to prepare a report of your company's activity as a sample. (If your department is a smaller organization with but one company or station, your chief will be asking you to prepare this report for the entire department.) This will work for either a volunteer department or a small career department.

For example
   During the past year:

   Total number of calls

   Total fire calls

   Total explosions (with no fire)

   Total EMS calls

   Total HAZMAT calls

   Total service calls

   Total good intent calls

   Total false alarms or false calls

   Total calls for natural causes (severe weather, flood, etc.)

   Total special calls

   Total undetermined calls

For definitions and additional information, see NFPA 901, *Standard Classifications for Incident Reporting and Fire Protection Data,* or the appropriate modules of the Fire Incident Reporting System in your area.

Your assignment: Prepare a memo report for your instructor that includes the data requested. You should get and use actual data for your company or department. Make a diligent effort and be honest. Prepare the data in both tabular and chart form. The appropriate chart is a pie chart similar to Figure 3-4 on page 58 of the textbook. Include a cover memo that highlights any significant findings.

For Fire Officer II

Prepare a Report

What NFPA 1021 says: The Fire Officer II shall "prepare a concise report for transmittal to a supervisor, given fire department record(s) and a specific request for details such as trends, variances, or other related topics."

Situation:
Your are temporarily assigned to headquarters. The chief has asked you to prepare a report of the accomplishments of the entire department for the previous calendar year.

Your assignment: Prepare and submit to your instructor a memo report using your department's actual records to show the department's activities including emergency calls, training, pre-incident planning, and inspection activities, fire and life-safety activities, and so on. You may want to copy and use the form provided in this manual for this purpose.

CHAPTER QUIZ

1. The first step in the management process is:
   a. commanding
   b. coordinating
   c. organizing
   d. planning

2. The need for _____ skills increases as one moves up the organizational ladder.
   a. command
   b. management
   c. planning
   d. technical

3. The fourth management function involves the _____ of available resources.
   a. commanding
   b. coordinating
   c. management
   d. organizing

4. The functions of management include:
   a. planning, organizing, commanding, coordinating, and controlling
   b. research, organizing, executing, coordinating, and leadership
   c. executing, planning, revising, coordinating, and controlling
   d. planning, organizing, controlling, and evaluating

5. _____ is the breaking down of large tasks into smaller ones.
   a. commanding
   b. controlling
   c. coordinating
   d. organizing

6. Management can be enhanced through good _____, _____, and personal observation.
   a. leadership, motivation
   b. communication, structure
   c. policy, procedures,
   d. leadership, policies

7. A system of conduct, principle, of honor, and morality describe:
   a. ethics
   b. human rights
   c. loyalty
   d. management

8. _____ is a principle that allows horizontal communication within the vertical command structure.
   a. customer service
   b. Theory Z
   c. Fayol's bridge
   d. the principle of exception

9. Which best describes the time frame that mid-range planning encompasses?

   a. within 1 year

   b. 1 to 5 years

   c. 5 to 10 years

   d. beyond 10 years

10. The term *internal customers* refers to:

    a. the public

    b. the local government

    c. the business owners

    d. the employees

11. Ethics begins where _____ leaves off.

    a. communications

    b. delegation

    c. justice

    d. the law

12. The management philosophy that involves a participatory style of leadership in which employees can be self-directed and handle authority is known as:

    a. Theory A

    b. Theory B

    c. Theory X

    d. Theory Y

13. An ethical leader:

    a. operates with integrity

    b. considers the impact of his or her actions on others

    c. is guided by a conscience

    d. all of the above

CHAPTER 5

THE COMPANY OFFICER'S ROLE IN MANAGING RESOURCES

OUTLINE

- Objectives
- Overview
- Ten Proven Time Eaters
- What Are Your Personal Time Thieves
- Assess the Needs of Your Customers

- Building a Management Team
- Ongoing and Emerging Management Issues for the Fire Service
- Ongoing Issues of National Importance
- Emerging Issues of National Importance

- Key Terms
- Review Questions
- Discussion Questions
- Homework
- Chapter Quiz

OBJECTIVES

Upon completion of this chapter, you should be able to describe:
- The importance of organizational vision and mission statements
- How goals and objectives support the department's mission
- How to effectively manage human resources, time, and the department's financial resources
- How to introduce and manage change
- A process approach when solving problems
- The role of the customer

OVERVIEW

This chapter examines everything covered in the book so far: The company officer's role, how the organization works, how management helps the organization accomplish its assigned mission, and how we communicate effectively. This chapter also emphasizes the practical application of helping you effectively manage your resources, especially your own time. As you get older, you will find that it is the most precious resource you have.

This chapter further recognizes the real reason for your organization's existence—the customers. Customers are important. You are entrusted with a mission to respond when they need help. In professional organizations, members are paid to maintain personal readiness and courteously provide services. In all organizations, members are entrusted with equipment. Part of your job is to maintain all equipment with a professional and personal labor of love (see **Figure 5-1**).

Figure 5-1 Washing the firetruck may seem like a chore for some, but others see it as an essential part of maintaining equipment that is ready and safe. Remember that the equipment belongs to someone else, in this case the taxpayers.

TEN PROVEN TIME EATERS

1. **Lack of personal goals and objectives.** Where are you headed?
2. **Lack of planning.** Set priorities, set a schedule to meet your goals, and follow through on your priorities. Manage your time; do not let time manage you.
3. **Procrastination** is putting off things that need to be done. It is poor time management. Making the decision to put off the activity takes time. It is often more effective to just bite the bullet and do it.
4. **Reacting to urgent events,** often the result of procrastination. These events force you to rearrange your schedule to meet the needs of the organization or other individuals.
5. **Telephone interruptions.** They are frequently unwanted and frequently unnecessary. Learn to conduct business in a polite, businesslike way and get off the phone.
6. **Drop-in visitors.** Be polite, do your business, and move along.
7. **Trying to do too much yourself.** Learn to focus on what is important and learn to delegate what you can.
8. **Ineffective delegation.** When delegating, learn the steps for doing it effectively.
9. **Personal disorganization.** In its milder forms, disorganization presents problems due to scheduling conflicts, the inability to find things when needed, and so forth. In a more serious form, it means you have not set goals and objectives and that you lack focus.
10. **Inability to say "no."** For some, this item may be the hardest on the list. Tie your time to your goals. Focus on priorities. If the request supports your goals, go for it. If it does not, politely say that you are sorry, but you are unable to take on that activity. Obviously this may not set well with your supervisor, but even here, supervisors can sometimes overload your plate. Show your supervisor a list of projects you are working on and ask for help in setting priorities. Guidance can be given and your supervisor may find, to his or her surprise, that you are, indeed, overloaded.

WHAT ARE YOUR PERSONAL TIME THIEVES?

Rank your five leading time thieves from each list. See **Figure 5-2.**

Personal

_____   Lack of priorities

_____   Lack of planning

Figure 5-2 Time management is important. Use your teammates to plan and organize your company's time effectively.

_____ Management by crisis

_____ Lack of delegation

_____ Overcontrolling subordinates

_____ Excessive dealings with routine and trivia

_____ Too much paperwork

_____ Too much reading

_____ Lack of procedure for routine matters

_____ Too much duplication of effort

_____ Other _____

Environmental

_____ Too many drop-in visitors

_____ Too many meetings

_____ Too many phone calls

_____ Too much organizational red tape

_____ Too many disorganized files

_____ Too many requests from subordinates

_____ Other _____

Solving Problems

When you have a problem, you should follow a proven, logical approach to find the solution. The first step may simply be to recognize that you *have* a problem. Many leaders do not see problems occurring, or if they do, they fail to take any corrective action. Here are the generally accepted steps for problem solving:

1. **Accurately define the problem.** This may seem simple, but in many cases, it the most difficult step. Also in many cases, it is a critical step, for if you do not correctly identify the problem, the rest of the process is doomed.

2. **Gather information.** Review the laws, policy statements, and regulations from federal, state, and local sources that may address the issue.

3. **Analyze the information.** See if you have everything you need.

4. **Look for alternative solutions** and list all possible solutions. At this point in the process, you want to be open minded. Invite others to offer ideas.

5. **Select one or more of the best solutions.** Rank the solutions in terms of expected outcomes. Some solutions may cost too much or take too long. There are always compromises. If the answers were easy, you would not be in this process. Pick one that you think best meets all of your objectives.

6. **Take action.** Now you must get approval. You have done all the work—identified the problem, gathered data, considered alternative solutions, and listed one or more recommended solutions. This information is usually conveyed to the supervisor in a short memo. Your supervisor should be able to quickly review what you have done and give approval.

7. **Monitor the results.** This step is important but often overlooked. Be sure you follow the process and see what happens. Most changes bring about some negative consequences. Also be sure the good consequences outweigh any bad items you introduce.

8. **Take corrective action if necessary.** Here you may need to reboot the process and start all over.

Most supervisors prefer to have a specific recommendation on a course of action to take. For example, in the case of the complaint about the outside speaker noise ("A Citizen's Complaint," p. 26), most chiefs prefer your recommendation for a clear course of action rather than passing the decision up to them. You may want to list several alternatives, but you should select one and support it with a logical argument. If you have done a good job in the problem-solving process, the logic of your choice will be apparent. This approach makes the decision-making process easy for the chief, offering a "yes" or "no" vote to approve or disapprove your recommendation. Using this approach focuses the chief's consideration on your first-choice solution and saves time. You might even get some credit for the idea.

⊚ ASSESS THE NEEDS OF YOUR CUSTOMERS

1. Seek feedback from customers about their needs.
2. Seek feedback from customers about their degree of satisfaction.
3. Define requirements of the workplace to meet these needs.
4. Establish goals.
5. Define a management philosophy.
6. Develop a self-directed team.
7. Provide feedback to the team regarding performance.
8. Incorporate new ideas and technologies.
9. Seek to make continuing improvements to the process and the product.
10. Fine tune the mission.

⊚ BUILDING A MANAGEMENT TEAM

The following are characteristics of good management in a committee or group activity, including your company (see **Figure 5-3**):[1]

1. Provide an informal, comfortable, relaxed environment, a working atmosphere in which people are involved and interested.

---

[1]Adapted from McGregor, Douglas, *The Human Side of Enterprise,* New York: McGraw-Hill Book Company, 1985, pp. 227–235.

2. Provide an environment that encourages discussion in which virtually everyone participates.

3. Provide an environment in which the objective is well understood and accepted by the members.

4. Provide an environment in which the members listen to each other and the discussion does not jump from one idea to another. Every idea is heard, and people do not feel foolish by offering a creative thought, even if it seems extreme.

5. Provide an environment in which disagreement is allowed. The group should be comfortable with this concept and show no signs of having to avoid conflict.

6. Provide an environment in which agreements are not suppressed or overridden by premature group action. However, there should be no tyranny in the minority. Individuals who disagree should not dominate the group or express hostility. Their disagreement should be a genuine difference of opinion, and they can expect a hearing to find a solution.

7. Provide an environment in which most decisions are reached by consensus, and it is clear that everybody is in general agreement.

8. Provide an environment where criticism is allowed but where there is little evidence of personal attack, either openly or in a hidden fashion. The criticism has a constructive flavor in that it is oriented toward removing an obstacle that faces the group and prevents it from getting the job done.

9. Provide an environment where people are free in expressing their feelings as well as their ideas on a problem and on the group's operation.

10. Provide an environment where clear assignments are made and accepted.

11. As the leader, provide an environment where you lead but do not dominate. In fact, allow the leadership to shift from time to time, depending on the circumstances. Different members, because of their knowledge or experience, are in a position at various times to act as resources for the group. The members utilize themselves in this fashion, and they occupy leadership roles while they are thus being used.

The traditional pyramid-shaped organizational chart shown in the text may not reflect your organization if you really value your members and respect your customers. In the traditional chart, the member is at the bottom and the customer is not even shown. If these two groups are as important as we say they are, and if the organization is really supportive of those who deliver the service listed in the organization's mission statement, then maybe the organization chart would look like the diagram shown in **Figure 5-4.**

Figure 5-3 Successful company officers create an environment where there is open discussion and agreement by consensus.

Figure 5-4 The traditional pyramid organization chart may need to be modified a bit to accurately reflect the focus in modern organizations on customer service and on the service provider. This illustration places customers at the top of the organization, supported by the service providers.

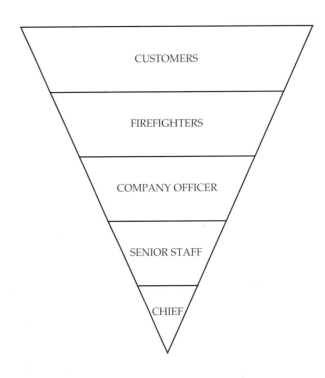

CUSTOMERS

FIREFIGHTERS

COMPANY OFFICER

SENIOR STAFF

CHIEF

ONGOING AND EMERGING MANAGEMENT ISSUES FOR THE FIRE SERVICE[2]

Before we close out this two-chapter discussion of management, we should look at where the fire and emergency services are headed and give some thought to the major issues that face management in these organizations. Your department may already be involved in one or more of these issues. As a member you may be involved in helping deal with some of these as well. Most of these topics are addressed in the book.

ONGOING ISSUES OF NATIONAL IMPORTANCE

▌ **Leadership.** To move successfully into the future, the fire service needs leaders capable of developing and managing their organizations in dramatically changed environments.

▌ **Prevention and public education.** The fire service must continue to expand the resources allocated to prevention, health, and safety education activities.

▌ **Training and education.** Fire service managers must increase their professional standing in order to remain credible to community policymakers and to the public. This professionalism should be grounded firmly in an integrated system of nationally recognized and/or certified education and training.

▌ **Fire and life-safety systems.** The fire service must support adoption of codes and standards that mandate the use of detection, alarm, and automatic fire sprinklers with special focus on residential properties.

▌ **Strategic partnerships.** The fire service must reach out to others to expand the circle of support to assure reaching the goals of public fire protection and other support activities.

▌ **Data.** To successfully measure service delivery and achievement of goals, the fire service must have relevant data and should support and participate in the revised National Fire Incident Reporting System. Likewise, NFIRS should provide the local fire service with relevant analysis of data collected.

▌ **Environmental issues.** The fire service must comply with the same federal, state, and local ordinances that apply to general industry, and which regulate response to and mitigation of incidents, plus personnel safety and training activities relating to the environment.

---

[2]Conference Report, Wingspread IV (1996). Distributed by the International Association of Fire Chiefs, Fairfax, Virginia. Used with permission of IAFC.

EMERGING ISSUES OF NATIONAL IMPORTANCE

■ **Customer service.** The fire service must broaden its focus from the traditional emphasis on suppression to a focus on discovering and meeting the needs of its customers. This may include preparing for and being able to deliver an effective response to all forms of local disasters, be they natural or man-made.

■ **Managed care.** Managed care may have the potential to reduce or control health care costs. It also will have a profound impact on the delivery and quality of emergency medical care.

■ **Competition and marketing.** In order to survive, the fire service must market itself and the service it provides, demonstrating to its customers the necessity and value of what it does.

■ **Service delivery.** The fire service must have a universally applicable standard that defines the functional organization, resources in terms of service objectives (types and level of service), operation, deployment, and evaluation of public fire protection and emergency medical services.

■ **Wellness.** The fire service must develop holistic wellness programs to ensure that firefighters are physically, mentally, and emotionally healthy and that they receive the support they need to remain healthy.

■ **Political realities.** Fire service organizations operate in local government arenas. Good labor/management and customer service relations are crucial to ensuring that fire departments have a maximum impact on decisions that affect their future.

Key Terms

Define in your own words the following terms:

agency shop _____

_____

arbitration _____

_____

authority _____

_____

budget _____

_____

capital budget _____

_____

closed shop _____

_____

fact finding _____

_____

goals _____

_____

line-item budget _____

_____

mediation _____

_____

mission statement _____

_____

objectives _____

_____

open shop _____

_____

operating budget _____

_____

program budget _____

_____

union shop _____

_____

vision _____

_____

QUESTIONS

Review Questions

1. What is a *mission statement*?

_____

_____

_____

_____

2. What is meant by *labor relations*?

_____

_____

_____

_____

3. What are some of the typical issues covered by union contracts?

_____
_____
_____
_____

4. What is a *budget*?

_____
_____
_____
_____

5. What is the difference between a *capital budget* and an *operating budget*?

_____
_____
_____
_____

6. How can you, as company officers, improve the use of the department's financial resources?

_____
_____
_____
_____

7. How can you, as company officers, improve the use of the department's human resources?

_____
_____
_____
_____

8. How can you, as company officers, improve the use of the department's time resources?

_____
_____
_____
_____

9. How can you, as company officers, improve the use of your personal time?

_____
_____
_____

10. What is your role, as a company officer, in managing customer service?

_____

_____

_____

_____

Discussion Questions

1. What are the benefits of a mission statement?

_____

_____

_____

_____

2. What is the mission statement of your organization? Do you agree with what it says? If you could make changes to your department's mission statement, what would you suggest?

_____

_____

_____

_____

3. Why is customer service an important issue for emergency response organizations?

_____

_____

_____

_____

4. Why is delegating difficult?

_____

_____

_____

_____

5. Why do some supervisors have problems delegating work?

_____

_____

_____

_____

6. Much of an officer's time is spent doing administrative work. What administrative duties would be appropriate for delegating? What administrative duties would *not* be appropriate for delegation?

_____

_____

_____

_____

7. How does your department manage routine expenditures?

_____

_____

_____

_____

8. How does your department manage capital expenditures?

_____

_____

_____

_____

9. How well do you manage your own time? What could you do to improve your time management?

_____

_____

_____

_____

10. What are the ongoing and emerging management issues in your department?

_____

_____

_____

_____

HOMEWORK FOR CHAPTER 5

For Fire Officer I

What NFPA 1021 says: Section 4.4.3. The Fire Officer I shall "prepare a budget request, given a need and budget forms, so that the request is in the proper format and is supported with data."

Situation:
With increasing demands for your company to prepare reports and make presentations for the community, you would like to purchase a laptop computer and a video projector. According to the department's policies, such items must be supported by justification and three bids.

Your assignment: Prepare a memo to your instructor justifying the need for the computer and projector. Get three actual bids for each item and attach them to your memo. If you have a preference as to which product to buy, address your preferences in your memo. Follow your department SOPs and use your department's forms as appropriate (see **Figure 5-5**).

BUDGET REQUEST

Department: _____ Division: _____

Person making request: _____

Date: _____

Description (Include make, model, serial numbers if appropriate. Attach reference material such as product sheet, copy of catalog page, etc., include encumbrance cost.):

Justification (Fully explain the benefit of the project—how it will make you more efficient, effective, or provide higher quality service. Explain how it will reduce costs or how on-going costs will be affected.):

Budget (Provide budget line-item or capital budget where budgeted during budget cycle. Include the amount budgeted. If not in approved budget, supply appropriate expenditure line and justification for approving an unbudgeted item and the affect on items budgeted. Line items up to $5,000, Department Capital to $5,000 to $10,000. All items over $10,000 are city capital budget items.):

Figure 5-5 Budget request form.

For Fire Officer II

What NFPA 1021 says: "This duty involves preparing a project or divisional budget, news releases, and policy changes, according to the following job performance requirements."
5.4.3. Describe the process of purchasing, including soliciting and awarding bids, given established specifications, in order to ensure competitive bidding.

Situation:

As your fire and life safety activities increase, it becomes apparent that a vehicle is needed for transportation for the educators.

Your assignment: Select an appropriate vehicle, such as a Ford Escort or similar. Go to three dealers that offer such vehicles. Get an honest bid for three similarly equipped vehicles. Submit the three bids and a cover memo indicating your needs, the process, and your recommendation regarding the purchase.

In most departments this purchase will be a capital expenditure. What special procedures are required for such procurements? Address this as part of your memo.

CHAPTER QUIZ

1. A _____ _____ declares the vision of the organization.
   a. clear direction
   b. mission statement
   c. written policy
   d. written procedure

2. The first step in problem solving is to:
   a. define the problem
   b. gather information
   c. recognize that there is a problem
   d. take action

3. The accepted steps in problem solving are:
   a. gather information
   b. define the problem
   c. look for alternative solutions
   d. all of the above

4. The major components of customer service include all of the following except:
   a. being nice to others
   b. profit margins
   c. raising the standard
   d. regarding everyone as a customer

5. _____ refers to a formal relationship between a group of members and the organization.
   a. union shop
   b. labor relations
   c. agency shop
   d. collective bargaining

6. Delegation can be defined as:
   a. giving others tasks and the authority and responsibility to do the task
   b. giving others tasks, but withholding authority and responsibility
   c. giving work to the best-qualified people
   d. having others do your work

7. A plan for using allotted funds to achieve the goals of the organization is called the:
   a. allotment
   b. budget
   c. cash flow
   d. funds source

8. Which of the following is managed by the company officer?
   a. time
   b. people
   c. budget
   d. all of the above

9. Which of the following is in the correct order, from broad to narrow in scope?
   a. mission statement, goals, objectives, tasks
   b. goals, objectives, tasks, mission statements
   c. objectives, tasks, mission statements, goals
   d. tasks, mission statements, goals, objectives

10. Barriers to delegation include the perceptions that:
   a. one can do it better themselves
   b. one is not willing to trust others
   c. one thinks that the time and effort required to train someone will take too long
   d. all of the above

CHAPTER 6

THE COMPANY OFFICER'S ROLE: PRINCIPLES OF LEADERSHIP

OUTLINE

- Objectives
- Overview
- Managers Versus Leaders
- Some Good Rules for Human Behavior
- Good Leadership

- What to Do When Harassment Occurs
- Redesigning Your Organization
- Key Terms

- Review Questions
- Discussion Questions
- Homework
- Chapter Quiz

OBJECTIVES

Upon completion of this chapter, you should be able to describe:

- Leadership
- The difference between leadership and management
- The leader's role and responsibilities in the organization
- Issues facing today's leaders

OVERVIEW

Up to this point, we have been looking at organizations and the management of resources. The most important resource in any organization is its people. To be effective, people working in organizations need good leadership. Chapters 6 and 7 in the textbook discuss the company officer's role as a leader. Leadership focuses on people and encourages them to work together. One of the most important jobs of the company officer is leadership because it involves working with others and trying to influence them to accomplish the organization's goals. This can be done in many ways, but when done effectively, productivity is increased and supervisors and workers seem happier and more satisfied (see **Figure 6-1**). Part of working with others involves accepting proper organizational behavior where everyone is treated fairly and with respect.

72

Figure 6-1 Effective leadership involves keeping your teammates motivated and interested in their work.

MANAGERS VERSUS LEADERS

1. Managers delegate work; leaders delegate responsibilities.
2. Managers tell people what to do; leaders explain why it has to be done.
3. Managers administer; leaders innovate.
4. Managers maintain; leaders develop.
5. Managers focus on systems; leaders focus on people.
6. Managers use controls; leaders use trust.
7. Managers have an eye on the bottom line; leaders have an eye on the horizon.
8. Managers do things right; leaders do the right thing.

SOME GOOD RULES FOR HUMAN BEHAVIOR

1. Accept people as they are.
2. Approach relationships and problems in terms of the present.
3. Treat people close to you with the same courtesy as those you just met.
4. Trust others.
5. Do without constant approval and recognition.[1]

---

[1]With thanks to Gary Briese, CAE, Executive Director, International Association of Fire Chiefs, Fairfax, Virginia.

GOOD LEADERSHIP

1. Focus on accomplishment.
2. Temper your lust to lead with good preparation.
3. Look for experience you can share with others.
4. Look for opportunities to improve the work of others.
5. Leadership success results from:
    a. Hard work that overcomes all forms of disappointment.
    b. Constantly carrying out the duties of one's office.
    c. Exercising the responsibility of leadership.
6. Never accept an offer of leadership if you are not willing to pay tributes necessary to successfully fulfill its obligations.

Power

How do you define power in the context of leadership? The word *power* is used several times in Chapter 6. *Leadership* is defined as achieving the organization's goals through others. To maintain this influence, the leader must have power over subordinate members. *Power* is the ability to influence others. You acquire leadership power in several ways.

Legitimate power. For fire service personnel, the first type of leadership power comes with the badge of the office. Some refer to this as *legitimate power,* the power that is bestowed upon you as an officer in the organization. As an officer, you wear insignia indicating that you are a representative of local government. Usually, the name of the governing agency is prominently displayed on that insignia. So, some power comes with the position.

Reward power. With legitimate power, two additional types of power are implied. The first of these is the ability to reward people, or *reward power.* As a supervisor, you have the power to approve requests, recommend individuals for special assignments, and write recommendations and evaluations that will help the members in your company attain their personal goals. Reward power is more than giving someone a medal or time off. It may be nothing more than name recognition, a smile, or an acknowledgment of effort. Reward power means taking a moment to help someone who is having a bad day. It does not mean taking your bad day out on someone else.

Punishment power. The other implied power that goes with the job is the power to punish, or *punishment power.* We usually think of punishment power as the authority to administer discipline, although, fortunately, this is not really an issue for most members. Far more often, we punish people by failing to give recognition for a job well done or withholding information that might be useful. These actions are not done on purpose. Instead, they represent failure on the part of the supervisor to do important things that define a good leader. Take a moment to look back at Herzburg's list of motivators.

Legitimate, reward, and punishment power come with the badge. There are, however, additional forms of power that one can earn. While legitimate, reward, and punishment powers come with the position, this added power is *earned* by individuals through their personal actions. It may be influenced by the way they treat people, by their ability to communicate, and by their knowledge of the job. These qualities can be illustrated with terms like *charisma* and *knowledge.* Clearly, these qualities are much more subjective than the legitimate power evidenced by the gold badge, but they are just as real. Terms such as *identification power* and *expert power* are often used to describe the qualities of role models and knowledgeable individuals.

Identification power. This is the recognition of authority by virtue of an individual's character or trust. Role models have identification power.

Expert power. This is the recognition of authority by virtue of an individual's skill or knowledge. Skill and knowledge are not synonymous with rank. Anyone with these qualities will have power in situations where they are useful in solving the problem at hand.

How to Determine the Leadership Style That Is Right for You

Effective leaders work at achieving the organization's goals through the efficient labor of others. They strive to do this as well as possible, and at the same time, help subordinates reach their full potential.

In real life, these goals are seldom attained easily. We are frequently faced with the old debate: Which is more important—the task or the people? Sometimes the task is critical. At other times, the people are far more important. Sometimes both are important; sometimes neither require much attention.

Elements That Determine Leadership Style

Three factors that can help you determine your own leadership style are the member, the leader, and the situation.

1. The Member
   - Experience
   - Maturity
   - Motivation
2. The Leader
   - Self-confidence
   - Confidence in the members
   - Feelings of security in the organization
   - Perception of the organization's value system
3. The Situation
   - Risk factors involved
   - Time constraints imposed
   - Nature of the particular problem
   - The organizational risk climate
   - The ability of individuals to work as a team

WHAT TO DO WHEN HARASSMENT OCCURS

Supervisors have a special obligation to protect members from the consequences of sexual harassment. When a member comes to you, a supervisor, with a complaint about sexual harassment, you must take action promptly. Encourage the complainant to provide you with specific information. You may need to help the process by asking open-ended questions like:

1. Where did this occur?
2. When did this occur?
3. Who was involved?
4. Were there any witnesses?
5. Have you talked to anyone else about this?
6. Has this happened before?
7. How long has this been happening?
8. Did you try to stop this conduct on your own?
9. What was the reaction?
10. What do you want me to do?

Some Guidelines on Asking Questions

1. When talking to witnesses, start by saying, "I am investigating a complaint of sexual harassment. I would like to ask you a few questions regarding what you may have seen or heard earlier today. . . ."
2. When talking to the accused, start by saying "I am investigating a complaint of sexual harassment by you. I would like to ask you a few questions regarding your actions or conversations this morning. . . ."
3. When talking to the accused, ask the individual to respond to the allegation(s). In most cases, the accused will acknowledge his or her actions and offer an excuse that the actions were misunderstood. Instruct the person on the seriousness of the accusation and advise him or her regarding what is and is not acceptable workplace conduct. Advise that there are laws and regulations to deal with these matters, and that violations can have serious consequences.

General Comments Regarding Harassment Issues

1. Move quickly.
2. Let everyone you talk to understand that you are concerned, that you take these matters seriously, and that you will make every effort to be fair and just.
3. Meet in private.
4. Make notes of the meetings and safeguard them.
5. Provide feedback to the accuser as to what you found.
6. The laws provide clear guidelines for organizations to follow in these matters. Know and follow your organization's policies.

🐸 REDESIGNING YOUR ORGANIZATION

In light of what we have learned about taking care of our people and our customers, the traditional organizational pyramid shown in the text may not accurately describe your organization (see Figure 1-3, p. 6 of the manual). If customers are the reason for our existence, they should be at the top of the chart. Next should be those who connect with and serve the customers (see Figure 5-2, p. 54 of the manual). What follows should support these customer-service providers.

Key Terms

Define in your own words the following terms:

consulting _____

_____

directing _____

_____

diversity _____

_____

EEOC _____

_____

expert power _____

_____

first-level supervisors _____

_____

harassment _____

_____

hygiene factors _____

_____

identification power _____

_____

leadership _____

_____

legitimate power_____

_____

model _____

_____

motivators _____

_____

organization _____

_____

power _____

_____

punishment power _____

_____

reward power _____

_____

supporting _____

_____

QUESTIONS

Review Questions

1. What is a group? What are some of the characteristics of a group?

_____

_____

_____

_____

2. Why do people like to work in groups?

_____

_____

_____

_____

3. What is *leadership*?

_____

_____

_____

_____

4. What is *leadership power*?

_____

_____

_____

_____

5. How does Maslow's theory apply to the leader's job in the fire service?

_____

_____

_____

_____

6. Why should the demographics of the fire department mirror those of the community it serves?

_____

_____

_____

_____

7. How does diversity help the fire service better serve its community?

_____

_____

_____

_____

8. Name and briefly describe the principal laws that provide guidelines to encourage diversity in the workplace.

_____

_____

_____

_____

9. Name and briefly describe the federal laws that prohibit sexual harassment in the workplace.

_____

_____

_____

_____

10. What action should you take, as a company officer, when a member comes to you with an allegation of sexual harassment?

_____

_____

_____

_____

Discussion Questions

1. Is leadership a learned or a natural skill?

_____

_____

_____

_____

2. Is there any benefit of studying the leadership traits of others?

_____

_____

_____

_____

3. How are leaders selected in your organization? Is this a valid process? Might there be a better solution?

_____

_____

_____

_____

4. Which leadership style is best for dealing with routine activity at the fire station?

_____

_____

_____

_____

5. Which leadership style is best for dealing with implementing a new procedure?

_____

_____

_____

_____

6. Which leadership style is best for dealing with the conditions at the scene of an emergency?

_____

_____

_____

_____

7. Discuss the five types of power discussed in the chapter.

_____

_____

_____

_____

8. Which type of power is most effective for company officers?

_____

_____

_____

_____

9. How does one determine the leadership style that is best for them?

_____

_____

_____

_____

10. The text suggests that diversity and harassment have been issues in the fire service. Would you agree? How did the fire service in your community respond to these issues? What has or is being done in the fire service in your community in response to these issues?

_____

_____

_____

_____

HOMEWORK FOR CHAPTER 6

For Fire Officer I

What NFPA 1021 says: "This duty involves general administrative functions and the implementation of departmental policies and procedures at the unit level, according to the following job performance requirements."

4.4.1. The Fire Officer I shall "recommend changes to existing departmental policies and/or implement a new departmental policy at the unit level, given a new departmental policy, so that the policy is communicated to and understood by unit members."

Carrying the High-Rise Pack

Situation:
Your department has an SOP for a high-rise hose pack. Some personnel have difficulty moving the pack outlined in the SOP. From your own experience, you know that the pack is heavy and awkward. In addition to problems in the pack itself, some personnel appear to lack the strength to deal with the pack.

These problems are not really significant at fires because there has not been a situation where the deployment of the pack had an outcome on the event. However, at drills and during training activity, the inability of some members to perform this task has created some bad feelings. Some firefighters have complained about having to do more than their fair share of the work. Some officers have complained to the chief about the department's recruit training. The training officer has complained about the physical standards the department set for recruits. The chief has indicated that he wants the problem taken care of, preferably at the company level.

The present high-rise pack was developed about 10 years ago. When initially adopted, the specifications addressed the contents of the pack—the nozzle, tools, and hose, as well as the design of the bag in which the items are carried. The Department's latest SOP appears on page 85 (**Figure 6-2**).

Your assignment: Prepare a memo addressed to your supervisor, with a copy to your instructor, fully outlining the problem as you see it and suggesting changes for solving the problem for the personnel in your company.

Your memo should describe how you arrived at the problem statement and how you developed your solution. Use your imagination but be realistic.

For Fire Officer II

What NFPA 1021 says: Section 5.4.2. "The Fire Officer II shall develop a project or divisional budget, given schedules and guidelines concerning its preparation, so that capital, operating, and personnel costs are determined and justified."

Your assignment: Develop an annual budget for your station. List all of the items that are routinely purchased and consumed during the course of the year. These will include housekeeping supplies, utilities, and other items for running the station. Include at least one item that needs replacement, such as a major kitchen appliance. Include three bids for a suitable replacement appliance.

CHAPTER QUIZ

1. A _____ is a group of people working together to accomplish a task.
    a. club
    b. department
    c. organization
    d. union

2. Leadership can be defined as:
    a. the actions needed to get a member to carry out certain tasks
    b. the position one holds
    c. the amount of power one has over personnel
    d. the number of subordinates one supervises

ANYTOWN FIRE DEPARTMENT
OPERATIONAL MANUAL
BOOK NUMBER 3
EMERGENCY OPERATIONS

| Chapter IV: | **FIRE FIGHTING** |
| Subject: | **HIGH RISE PACK** |

Code: **3- IV -13**   PAGE   1 OF 1   DATE 3/1/94

13-01 PURPOSE

A. To establish a procedure facilitating the most effective method for moving a high rise pack from a truck or engine company into a fire area so that the movement will be quick, as easy as is reasonable, and provide the most protection, from a safety standpoint, for our personnel.

B. To provide the proper tools needed for suppression activities in multistory buildings or any building where the high rise pack is deemed a necessary tool for the suppression of fire.

13.02 POLICY

To ensure that the proper equipment found in the high-rise pack be transported and used in building(s) where the high rise pack is needed as a suppression tool.

13.03 PROCEDURE

A. The movement of the high rise pack from street level to the floor or area designated by the Incident Manager will be the responsibility of the personnel of each individual company.

B. The high rise pack will be carried into a building where a standpipe system exists or where the distance is too great for a preconnected hose line.

C. The hose in the high rise pack shall be utilized in place of the hose line provided on the building standpipe system.

D. The pack may be moved in one of two ways in a multistory building.

    1. Via a stairwell.

    2. Via an elevator providing the elevator is fire department controlled.

13.04 SAFETY

No operation as outlined in this SOP shall preclude any person from using good judgment with due regard for the safety of all personnel.

Figure 6-2 SOP.

3. The leadership style in which a leader encourages participation and shares responsibility is called:

   a. directing

   b. consulting

   c. supporting

   d. delegating

4. The element that determines leadership style is:

   a. the leader

   b. the member

   c. the situation

   d. all of the above

5. Leaders can have several sources of power. Power that comes from charisma and the ability to make others wish to identify with the leader is known as:

   a. legitimate power

   b. expert power

   c. reward power

   d. identification power

6. Motivating members involves trying to do which of the following?

   a. getting them to move up Maslow's Hierarchy of Needs

   b. giving them more freedom

   c. getting them to do what you want

   d. satisfying their greatest need at the moment

7. Maslow's Hierarchy of Needs suggests that _____ needs are based on a person's desire to belong and be accepted.

   a. physiological

   b. security

   c. social

   d. self-esteem

8. The power tied to your position within an organization is referred to as _____ power.

   a. formal

   b. reward

   c. punishment

   d. expert

9. When is an organization liable for the sexually harassing actions of a member?

   a. when the organization knows about the situation and does nothing about it

   b. when the harasser is a supervisor

   c. when the organization knows, or should have known, about the situation

   d. all of the above

10. As defined by the EEOC, a hostile work environment involves:

   a. requests for sexual favors to gain promotion

   b. believing that men are chauvinists

   c. believing that women are not able to do a certain job

   d. treating employees differently because of their sex

THE COMPANY OFFICER'S ROLE IN LEADING OTHERS

 OUTLINE

- Objectives
- Overview
- What Kind of Person Are You?
- Tools for Effective Leaders

- Douglas McGregor's Hot Stove Rule
- Key Terms
- Review Questions

- Discussion Questions
- Homework
- Chapter Quiz

 OBJECTIVES

Upon completion of this chapter, you should be able to describe:

- How to help new teammates integrate into the organization
- How to help all teammates rise to their full potential
- How to understand and reduce conflict in your organization
- How to introduce change into your organization

 OVERVIEW

Leadership seems easy for some. However, the qualities of an effective leader are not always apparent. We see things happening but fail to realize the reason they are happening. In many cases, it is good leadership.

Many officers are placed in leadership positions before they are fully prepared. Part of becoming an effective leader must be learned through personal experience or what some call the "Road of Hard Knocks." It can be a rather difficult experience for the leader and the members who must work for that leader. Many of the tools of effective leadership can be learned through studying the teachings of others. An advantage of learning leadership in a training or educational environment is that it is less likely to affect the lives of others. Poor leadership in the workplace can lead to loss of productivity and morale, or possibly more serious consequences affecting the safety of firefighters and citizens.

 WHAT KIND OF PERSON ARE YOU?

There are several tests available to help you answer this question. Several, such as the Keirsey Temperament Sort, can be taken online. The Myers-Briggs test is another.

These tests present a series of questions that will help analyze your temperament, your personality, how you process information, and how you make decisions. Essentially your answers will lead to a score that will evaluate four dimensions of temperament as follows:

Extroversion	or	Introversion
Intuition	or	Sensing
Thinking	or	Feeling
Judgment	or	Perception

In each of these categories you are scored on a scale that ranges from 0 to 40. The greater the number indicated by your answers, the stronger the trait.

What can you do with this information? First, make sure you understand it yourself. Second, you can use it to better understand those around you. For example, if you know your boss's score, you can better understand and work with your boss. If you know your teammates' scores, you can better communicate with them, and use their respective skills to their fullest. It also works in improving relations at home.

For more information you may want to look at the Keirsey Temperament and charter Web site at http://keirsey. com/ or in a best-selling book entitled *Please Understand Me II*. See the list of additional resources at the end of Chapter 7 for more information.

TOOLS FOR EFFECTIVE LEADERS

Have an orientation session. We usually have an orientation session for new firefighters. What do we do for experienced firefighters who are transferred to a new station? What do we do when the supervisor changes? In every case, there is a need for taking a few minutes to get acquainted. For the supervisors who do this regularly, it is a good idea to have a short outline, similar to a teaching outline, to help facilitate the transition. This will help you cover every issue with every member. You may even want to make a copy of this outline for the member. This is a good time to explain personal values, as well as any policies, that may be unique to your workplace. This is also a good time to listen to the member and hear about his or her goals, ambitions, problems, and needs. Remember, you need not cover everything in one session. In most cases, you will find that the member will have better retention if the information is passed in several shorter sessions.

Have regular meetings and share information with members. One effective meeting that should happen regularly is a *planning meeting*. The planning meeting may be nothing more than getting together with everyone after the equipment checks at the start of the shift for a cup of coffee and a brief discussion about what will happen during the upcoming shift. Some meetings may be longer, such as planning the training schedule for the next three months. Plan for future events and let your members share the information. Let members be involved in the decision-making process whenever possible.

Set goals for the company and for every member. Some goals will be collaborative efforts that require group acceptance, whereas other goals may be for individual accomplishment. In any case, actively involve the members in the goal session activity. Use these goals as a benchmark for measuring progress during evaluations.

Provide regular feedback on performance. This does not require a formal evaluation process with a complicated report. It may only take a minute or two, and will usually focus on some specific task or time frame (see **Figure 7-1**). It is easier for the supervisor and the member to deal with performance-related issues when they are addressed immediately after the offending activity takes place. When corrective action is desired, it will come sooner.

Empower your people. Offer opportunities for your members to enhance themselves through growth and learning. Encourage them to seek training and education, and offer incentives for increasing their knowledge.

Provide instruction and help. Be available, show interest and support, and recognize accomplishments, but do not supervise unless necessary.

Figure 7-1 Counseling focuses on improving some specific aspect of member performance, attitude, or behavior.

Ask your members for ideas and suggestions. Listen to the members' ideas and suggestions. Respond to all of them and, when appropriate, see that the ideas are forwarded up the chain of command for action. Be sure that members get credit for their work and suggestions.

Be a role model. Do the right thing.

DOUGLAS MCGREGOR'S HOT STOVE RULE

Douglas McGregor, discussed in Chapter 4, offers some ideas for administering punishment that reduces resentment by the disciplined member.

The punishment should follow a warning. Small children are taught not to touch a hot stove. Those who fail to follow the advice will feel some pain. Members who fail to follow the rules will likewise feel some pain.

When one touches a hot stove, there is immediate feedback. For most, it is easy to connect the act with the consequences. Discipline should likewise be swift so that there is no misunderstanding about why the punishment was imposed.

The hot stove is impersonal and does not discriminate. All who touch it get burned. Likewise with punishment, supervisors should be impersonal and not discriminate when administering discipline.

In summary, supervisors should clearly let members know the rules. When the rules are broken, members should be swiftly disciplined in a way that is consistent and nondiscriminatory (see **Figure 7-2**).

Key Terms

Define in your own words the following terms:

coaching _____

_____

complaint _____

_____

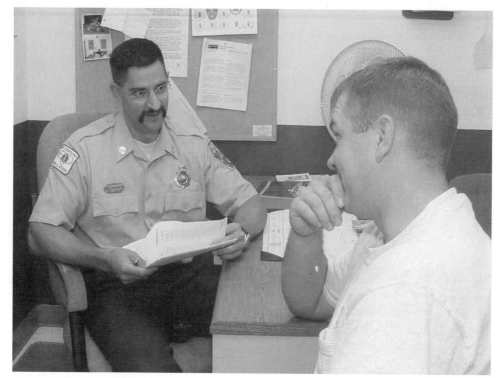

Figure 7-2 Company officers should let employees know when their work is not up to standard. Counseling and coaching provide opportunities to correct short-comings before they become problems.

conflict _____

_____

counseling _____

_____

demotion_____

_____

disciplinary action _____

_____

empowerment_____

_____

grievance _____

_____

grievance procedure _____

_____

gripe _____

_____

oral reprimand _____

_____

suspension _____

_____

termination _____

_____

transfer _____

_____

written reprimand _____

_____

QUESTIONS

Review Questions

1. What is *leadership*?

_____

_____

_____

_____

2. What is *coaching*?

_____

_____

_____

_____

3. What is *counseling*?

_____

_____

_____

_____

4. The text suggests that the performance evaluation system is a *continuous* process. Explain.

_____

_____

_____

5. Describe the disciplinary process in your department.

_____

_____

_____

_____

6. What are some general rules for effective punishment?

_____

_____

_____

_____

7. How should you, as company officers, deal with conflict?

_____

_____

_____

_____

8. How should you respond to complaints and grievances?

_____

_____

_____

_____

9. How should you introduce change?

_____

_____

_____

_____

10. How can you, as a company officer, become an effective leader?

_____

_____

_____

_____

Discussion Questions

1. As a company officer, describe how you would treat a new firefighter joining your company.

_____

_____

_____

_____

2. Describe how you would integrate this new firefighter into your team.

_____

_____

_____

_____

3. Describe how you would encourage the new firefighter's professional development.

_____

_____

_____

_____

4. Describe something you would do at your fire station to make it more attractive to the first minority member to join your company.

_____

_____

_____

_____

5. As a company officer, how would you introduce a new procedure (or a change to an existing one) mandated by law or regulation?

_____

_____

_____

_____

6. As a company officer, how would you introduce a new procedure (or a change to an existing one) that is the result of your own initiative?

_____

_____

_____

_____

7. As a company officer, how would you deal with a new idea suggested by one of your teammates?

_____

_____

_____

_____

8. How is change accepted in your department?

_____

_____

_____

_____

9. Other than those mentioned in the text and in this *Student Manual,* what changes are occurring in our industry?

_____

_____

_____

_____

10. What changes do you foresee in our business over the next five years?

_____

_____

_____

_____

HOMEWORK FOR CHAPTER 7

For Fire Officer I

What NFPA 1021 says: The Fire Officer I shall:

1. "Assign tasks or responsibilities to unit members, given an assignment under nonemergency conditions, such as at a station or other work location, so that the instructions are complete, clear, and concise; safety considerations are addressed; and the desired outcomes are conveyed.

2. Direct unit members during a training evolution, given a company training evolution and training policies and procedures, so that the evolution is performed in accordance with safety plans, efficiently, and as directed.

3. Assign tasks or responsibilities to unit members, given an assignment at an emergency operation, so that the instructions are complete, clear, and concise; safety considerations are addressed; and the desired outcomes are conveyed."

Your assignment: For each of the three requirements previously described, explain how you would assign duties and direct the activities of your teammates. In each case, you are being observed and evaluated by your supervisor, so you want to do a good job.

Your instructor may provide specific situations for each of the three required activities. This activity can be done orally or in writing at the discretion of your instructor.

For Fire Officer II

What NFPA 1021 says:

1. The Fire Officer II shall "initiate actions to maximize member performance and/or to correct unacceptable performance, given human resource policies and procedures, so that member and/or unit performance improves or the issue is referred to the next level of supervision."

2. In addition, the Fire Officer II shall "evaluate the job performance of assigned members, given personnel records and evaluation forms, so each member's performance is evaluated accurately and reported according to human resource policies and procedures."

Your assignment: Utilizing the policies and forms of your organization, describe your actions for each of the required activities. Your instructor may provide specific situations for each of the three situations previously listed. This activity can be done orally or in writing at the discretion of your instructor.

CHAPTER QUIZ

1. Fire officers who pitch in and personally help out when things are rushed should:
   a. be considered team players
   b. find that their subordinates will expect help all the time
   c. justify their action
   d. show respect of their subordinates

2. The most important rule to remember when introducing change is:
   a. be as forceful as necessary to "sell" the change
   b. discuss the change as early as possible
   c. downplay the methods that lead to the change
   d. follow-up on complaints that members may raise

3. What is the purpose of disciplinary action?

    a. to point out to members who is in charge

    b. to get even with a difficult member

    c. to change the behavior of a member

    d. to humiliate members in front of their peers

4. Which of the following penalties is most likely to be used by a company officer?

    a. demotion

    b. dismissal

    c. reprimand

    d. suspension

5. When should a supervisor give a member feedback on his or her work?

    a. only when directed by higher officers

    b. only during the performance appraisal interview

    c. only when the member requests it

    d. anytime when it is appropriate

6. Who is most likely to know the best way to improve on how a job can be done?

    a. the people who do the work

    b. the supervisor of the work team

    c. top management

    d. the oldest and most experienced member of the organization

7. The performance appraisal interview should include:

    a. a discussion of the effects of rater bias

    b. written criteria upon which performance is based

    c. a plan of action and goals for the coming year

    d. a complete position description

8. Three effective tools for member development are:

    a. education, opportunity, and advancement

    b. coaching, counseling, and evaluations

    c. education, coaching, and evaluations

    d. coaching, opportunities, and evaluations

9. Someone who helps another develop a skill set is called a(n):

    a. counselor

    b. coach

    c. evaluator

    d. teacher

10. Getting members involved is called:

    a. empowerment

    b. arbitration

    c. coaching

    d. participation

THE COMPANY OFFICER'S ROLE IN PERSONNEL SAFETY

OUTLINE

- Objectives
- Overview
- Ten Suggestions to Help Company Officers Improve Their Firefighters' Safety

- The Public Safety Officers' Benefits (PSOB) Act
- Key Terms
- Review Questions

- Discussion Questions
- Homework
- Chapter Quiz

OBJECTIVES

Upon completion of this chapter, you should be able to describe:

- The need for an ongoing concern for safety in the fire service
- The common causes of accidents and injuries in the fire service
- The signs and symptoms of stress among personnel
- How to develop and implement a safety program at the station level

OVERVIEW

While individual firefighters have a responsibility for their own well-being, company officers also have a moral and legal responsibility for the safety and survivability of their firefighters. If an accident occurs, company officers must explain their action, or the lack thereof.

NFPA 1500 compressively addresses a fire department health and safety program. When fire departments implement the procedures outlined in the standard, they will see improvements in the health and safety of their members. Many departments have already undertaken these steps and the results are already apparent, as there has been a significant reduction in the number of deaths in the fire service.

Providing a safe and healthful workplace should be the goal of every employer. The results of such an approach usually mean fewer accidents, higher morale, more productivity, and reduced liability resulting from accidents (see **Figure 8-1**). This proactive approach can result in reduced risk to the members of the fire service as well as a reduced risk to their departments. Many fire departments tend to be reactive, rather than proactive, when it comes to firefighter safety. This position is unfortunate for all.

Figure 8-1 NFPA 1500 has been around for more than 15 years and will be involved in any legal issue involving fire department safety practices. *Photo courtesy of the International Society of Fire Service Instructors.*

Regardless of the fire department's "organizational attitude" about safety, company officers can have a significant impact on their own and their subordinates' health and safety. Many of the requirements in NFPA 1500 can be introduced at the company level. As with many of the topics discussed in this book, setting the tone and enforcing the department's policies regarding safety is one of the company officer's many responsibilities. During emergency response, whether the call be a single-unit response for a dumpster or a three-alarm event, the company officer is responsible for the safety of their crew. At the scene of an emergency as well as in the fire station, firefighter safety is ultimately the company officer's responsibility.

The greatest risk (40–50 percent) is heart attack. A majority of these were the result of physical stress or overexertion, and many firefighters involved had records of heart problems. Age was also a factor in many of these situations. The statistics show that the probability of dying while on duty roughly doubles for each decade of life after 40.

The second leading cause of fatalities is motor-vehicle-related accidents. Excessive speed, failure to wear seat belts, and failure to yield at intersections contributed to many of the accidents.

Motor-vehicle accidents are not unique to the fire service. The National Institute for Occupational Health and Safety (NIOSH) recently reported that motor-vehicle accidents were a leading cause of fatalities in the workplace, noting that effective driver training programs and the use of seat belts can help significantly. NIOSH found that companies investing in driver training have a significantly lower incidence of accidents. NIOSH also noted that over half of the drivers involved in fatal accidents were NOT wearing their seat belt at the time of the accident.

Using seat belts, driving carefully, and following your organization's policies are human-behavior factors, not design problems or mechanical failures. They clearly reinforce the concept that safety is a personal attitude. Safer vehicles have reduced the number of fatalities in accidents involving firefighters falling from apparatuses. The availability and use of safer, personal protective equipment has improved conditions on the fireground (see **Figures 8-2a, 8-2b** and **8-2c**), but personal behavior continues to be a significant problem. Firefighters should be held responsible for their own safety. At the same time, company officers should recognize when unsafe conditions occur and take appropriate corrective action.

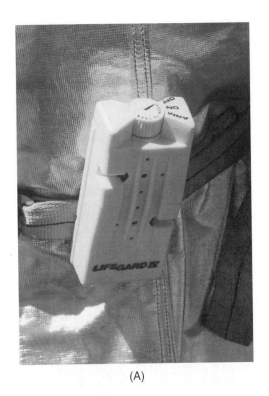

(A)

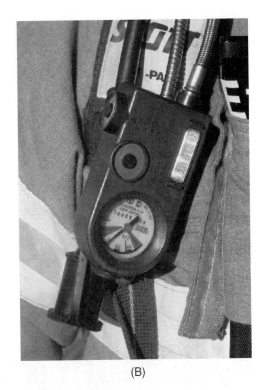

(B)

(C)

Figure 8-2 PASS devices can help save a firefighter's life—but they *must* be activated. (A) A manual PASS device *must* be armed by the wearer. (B and C) An integrated PASS device on a SCBA, which is activated when the wearer opens the SCBA bottle.

TEN SUGGESTIONS TO HELP COMPANY OFFICERS IMPROVE THEIR FIREFIGHTERS' SAFETY

1. Know the job. Company officers should constantly be working on improving their knowledge and skills.
2. Be observant of the actions of others.
3. Be a team player. Build team spirit and interdependence among all of the players.
4. Set a good personal example. Do not simply talk the talk; walk the walk as well.
5. Be obviously concerned about safety at all times.
6. Provide constructive feedback to firefighters.
7. Communicate effectively in all your dealings, but especially in the area of safety.
8. Be assertive when necessary! Sometimes it is the only way you can get compliance. This is not a time to be timid about your feelings.
9. When all else fails, feel free to take appropriate disciplinary action.
10. Be truly positive about safety and health issues.

THE PUBLIC SAFETY OFFICERS' BENEFITS (PSOB) ACT

The Public Safety Officers' Benefits (PSOB) Act (42 U.S.C. 3796, et seq.) was enacted in 1976 to assist in the recruitment and retention of law enforcement officers and firefighters. Specifically, congress was concerned that the hazards inherent in law enforcement and fire suppression and the low level of state and local death benefits might discourage qualified individuals from seeking careers in public safety, thus hindering the ability of communities to provide for public safety.

The PSOB Act was designed to offer peace of mind to men and women seeking careers in public safety and to make a strong statement about the value that American society places on the contributions of those who serve their communities in potentially dangerous circumstances.

The PSOB Program provides death benefits in the form of a one-time financial payment to the eligible survivors of public safety officers whose deaths are the direct and proximate result of a traumatic injury sustained in the line of duty. Since October 15, 1988, the benefit has been adjusted each year on October 1 to reflect the percentage of change in the Consumer Price Index. As of October 1, 2003, the benefit amount is $267,494. For each death and disability claim, the award amount is solely determined by the actual date of the officer's death or disability.

The PSOB Program provides disability benefits for public safety officers who have been permanently and totally disabled by a catastrophic personal injury sustained in the line of duty if that injury permanently prevents the officer from performing **any** substantial and gainful work. Medical retirement for a line-of-duty disability does not, in and of itself, establish eligibility for PSOB benefits.

The PSOB Program includes the Public Safety Officers' Educational Assistance (PSOEA) Act. This act expands on the former Federal Law Enforcement Dependents Assistance Program to provide financial assistance for higher education for the spouses and children of federal, state, and local public safety officers who have been permanently disabled or killed in the line of duty. Educational assistance through the PSOEA Program is only available to the spouse or children of a public safety officer after the PSOB death or disability claim process has been completed and benefits have been awarded. The educational assistance may be used to defray relevant expenses, including tuition and fees, room and board, books, supplies, and other education-related costs. As of October 1, 2003, the maximum award for a full-time student is $695 per month of class attendance. All PSOEA awards must, by law, be reduced by the amount of other governmental assistance that a student is eligible to receive.

As defined by Congress in Public Law 90-351 (sec. 1217), a public safety officer is an individual serving a public agency in an official capacity, with or without compensation, as a law enforcement officer, firefighter, or member of a rescue squad or ambulance crew. In October 2000, Public Law 106-390 (Sec. 305) designated members of the Federal Emergency Management Agency (FEMA) as public safety officers under the PSOB Act if they were performing official, hazardous duties related to a declared major disaster or emergency. The legislation also indicated that state, local, or tribal emergency management or civil defense agency members working in cooperation with FEMA are, under the same circumstances, considered public safety officers under the PSOB Act. Retroactive to September 11, 2001, chaplains also are included in the PSOB Act definition of a public safety officer.

The PSOB office works with national-level police and firefighter groups to provide visibility and emotional support to this unique constituency. Concerns of Police Survivors (COPS), Inc., provides services and assistance for families and coworkers of fallen law enforcement officers during the annual National Police Week program. They also provide regional training sessions and several special seminars and extended programs for spouses, siblings, and children of slain officers. The National Fallen Firefighters Foundation provides peer counseling, training, and technical assistance for the families and coworkers of firefighters who were killed in the line of duty. Some of their specific activities include developing and disseminating publications and reference materials for survivors and senior fire department managers and creating a public awareness strategy to promote the fire service and its critical role in public safety.

Death Benefits

State and local law enforcement officers and firefighters are covered for line-of-duty deaths occurring on or after September 29, 1976. Federal law enforcement officers and firefighters are covered for line-of-duty deaths occurring on or after October 12, 1984. Federal, state, and local public rescue squads and ambulance crews are covered for line-of-duty deaths occurring on or after October 15, 1986. As of October 30, 2000, employees of the Federal Emergency Management Agency (FEMA) and state, local, and tribal emergency management and civil defense agency employees working in cooperation with FEMA are considered to be public safety officers under the PSOB Program, provided they were performing official, hazardous duties related to a declared major disaster or emergency.

Disability Benefits

On November 29, 1990, the Act was amended to include public safety officers who became permanently and totally disabled as a result of a catastrophic injury. Public safety officers (federal, state, and local law enforcement officers, firefighters, and members of public rescue squads and ambulance crews) are covered for catastrophic personal injuries sustained on or after November 29, 1990. To initiate a claim for PSOB disability benefits, the officer must be separated from his or her agency for medical reasons and must be receiving the maximum allowable disability compensation from his or her jurisdiction. Eligible officers may include those who are comatose, in a persistent vegetative state, or quadriplegic.

Education Assistance

The Federal Law Enforcement Dependents Assistance (FLEDA) Act (Public Law 104-238 (PDF or ASCII)) was enacted in October 1996 to provide financial assistance for higher education to spouses and children of federal law enforcement officers killed in the line of duty. Congress and the President recognized that for many families, access to higher education was instrumental in their ability to move forward in the aftermath of a line-of-duty tragedy. The Act was amended in 1998 (Public Law 105-390 (PDF or ASCII)) to also provide educational assistance for spouses and children of state and local police, fire, and emergency public safety officers killed in the line of duty, creating the Public Safety Officers' Educational Assistance (PSOEA) Program. This program also makes assistance available to spouses and children of public safety officers who have been permanently and totally disabled by catastrophic injuries sustained in the line of duty. The Act was amended in October 2000 to extend the retroactive eligibility dates for financial assistance for higher education for spouses and dependent children of federal, state, and local law enforcement officers who were killed in the line of duty on or after January 1, 1978.

Applying for Death Benefits

Eligible survivors may file claims directly with the PSOB Office or through the public agency in which the public safety officer served.

1. After a fatality occurs, the department should:
   ▮ Make arrangements for an autopsy, which often provides the PSOB Office with useful information regarding the cause of death.
   ▮ Identify a department member to serve as a liaison between the department and the PSOB Office.

2. After being named, the department's liaison should:
   - ▌Call the PSOB Office at 1-888-744-6513.
   - ▌Provide accurate, up-to-date information regarding:
     - ☐ The public agency's name.
     - ☐ The liaison's name.
     - ☐ Phone numbers for the department and the liaison.
     - ☐ A fax number or mailing address so the PSOB Office can send the claim initiation guidance letter.
     - ☐ The name of the deceased (public safety officer).
     - ☐ The date of the incident and the deceased's date of death.
     - ☐ A brief description of the incident.
   - ▌Relay the information very carefully and include only what is known. There should be no speculation as to the cause of death.
   - ▌Leave a phone message with the liaison's name and telephone number if calling during the evening or on a weekend.
3. After being informed of an incident by the department liaison or other sources, the PSOB Office will mail a Claims Guidance Package to the department liaison as soon as possible.

Something to Remember

- ▌It takes a minute to write a safe rule.
- ▌It takes one hour to hold a safety meeting.
- ▌It takes one week to plan a safety program.
- ▌It takes one month to put the plan into full operation.
- ▌It takes one year to win the chief's safety award.
- ▌It takes just a second to destroy it all with an accident!

Key Terms

Define in your own words the following terms:

HSO _____

_____

personnel accountability_____

_____

rehabilitation _____

_____

RIT _____

_____

QUESTIONS
Review Questions

1. About how many firefighter injuries and fatalities occur each year?

2. Where do most firefighter injuries and fatalities occur?

3. List the most common causes of firefighter fatalities.

4. List the most common causes of firefighter injuries.

5. Under what conditions do these injuries occur?

6. What are some of the conditions that bring stress to firefighters? What are the signs and symptoms of stress?

7. What actions can you take as company officers to reduce injuries at the fire station?

_____

_____

_____

_____

8. What action can you take to reduce injuries while responding to emergencies?

_____

_____

_____

_____

9. What actions can you take to reduce injuries at the scene of emergencies?

_____

_____

_____

_____

10. What is your department's commitment to firefighter safety and health?

_____

_____

_____

_____

Discussion Questions

1. What can you do as company officers to enhance the safety of your personnel?

_____

_____

_____

_____

2. What significant changes have occurred in the fire service within the last 10 years as a result of health and safety concerns?

_____

_____

_____

_____

3. Based on your own observations, what causes firefighters to get injured?

_____

_____

_____

_____

4. If you were assigned duties as the incident safety officer at the scene of a working structure fire in a residence, what would be your concerns? What would be your duties?

_____

_____

_____

_____

5. What are some of the signs and symptoms of stress? What role do you have as a company officer in providing help for teammates showing signs or symptoms of stress?

_____

_____

_____

_____

6. What steps can you take to maintain a healthier body?

_____

_____

_____

_____

7. What steps can you take to maintain a healthier mind?

_____

_____

_____

_____

8. What steps can be taken by your company to improve firefighter safety and health?

_____

_____

_____

_____

9. What steps can be taken by your department to improve firefighter safety and health?

_____

_____

_____

_____

HOMEWORK FOR CHAPTER 8

For Fire Officer I

What NFPA 1021 says:

Section 4.7 "This duty involves integrating safety plans, policies, and procedures into the daily activities as well as the emergency scene, including the donning of appropriate levels of personal protective equipment to ensure a work environment, in accordance with health and safety plans, for all assigned members, according to the following job performance requirements (see **Figures 8-3a** and **8-3b**)."

Specifically, Section 4.7.2 says that the Fire Officer I shall "apply safety regulations at the unit level, given safety policies and procedures, so that required reports are completed, in-service training is conducted, and member responsibilities are conveyed."

Additionally, Section 4.7.2 says that the Fire Officer I shall also "conduct an initial accident investigation, given an incident and investigation form, so that the incident is documented and reports are processed in accordance with policies and procedures."

Situation:

As the company officer, you are expected to be a leader in dealing with teammates, supervisors, and the public. Part of your job includes a responsibility for you and your team performing in a safe manner that meets the expectations of these groups. As part of that job, you are also expected to take the initiative in providing opportunities to train your personnel.

Last Sunday, you took advantage of the nice weather and a rather slow day to facilitate driver training for firefighter Elmore Smart. Others in your crew that day were firefighters Schecky Lorraine and Phil Niemann. While on driver training, your route took you past the new Regal Burger Palace at 303 Tragic Avenue, where you observed construction on the new restaurant coming along with the walls up and a roof on the building.

Stopping about 3:00 P.M. to take a better look, you and your crew wandered around the site for a few minutes. While you started out together, the four of you divided up into pairs, and eventually each of you just sort of walked around, looking at items of interest.

You were just about done with the visit when you heard a crash and a scream. Upon investigation you found firefighter Schecky Lorraine at the bottom of the short stairway leading to a below-grade basement door. He had apparently walked across the stairway opening on a sheet of plywood, not realizing that it was unsupported. After determining the nature of his injuries, you called for an ambulance.

Lorraine was transported to the Community Hospital where he is being held for observation of complications from head, neck, and back injuries sustained in the fall. The doctor at the hospital said that Lorraine will be okay but estimated that he will be off duty for as much as two weeks.

After returning to headquarters to fill out the accident and injury reports you had a phone call from a Mr. Charles "Big Red" Ceder, the construction foreman from the Regal Burger project. He wanted to know what you were doing at the construction site and why you were there without permission or an escort. He said that he had heard of the accident (to Lorraine) and indicated that if the fire department took any legal action or if he or his company were cited for any safety violations, his company would sue you and your department for trespassing.

(A)   (B)

Figure 8-3 Duty personnel should set up their gear for rapid—and complete—donning. Establish good habits to help eliminate shortcuts.

Your assignment: Prepare a memo to your fire chief outlining the events of the day. Report what happened, describe the problems that have been created, and describe what actions you have taken to prevent such an event from reoccurring as well as your recommendations on how to prevent such events form occurring to others in the future.

Complete and attach the appropriate injury and accident forms, using your department's forms if available. If no forms are available use the forms provided here (**Figures 8-4** and **8-5**). Make up personal data where needed. Submit the memo and attachments to your instructor.

For Fire Officer II

Firefighter Health and Safety

What NFPA 1021 says:

Section 5.7 "This duty involves reviewing injury, accident, and health exposure reports, identifying unsafe work environments or behaviors, and taking approved action to prevent reoccurrence, according to the following job requirements."

Specifically Section 5.7.1 says that the Fire Officer II shall "analyze a member's accident, injury, or health exposure history, given a case study, so that a report including action taken and recommendations made is prepared for a supervisor."

Situation:

The chronically injured firefighter

One of your firefighters seems to have a propensity for getting injured. Most of his injuries are not serious, but they put him out of action. As a result, other firefighters have to do his work. It's becoming a problem for your company and for you personally.

Firefighter Kenneth L. Klutz has just returned from two weeks of sick leave. Two weeks ago, during a routine assist on an EMS call, he and the other members of your team were helping the ambulance crew move a patient from her home to the ambulance. To make this move they had to go down a flight of steps from the front porch to the sidewalk. During the move, Klutz was on the lower end of the stretcher. Upon returning to headquarters, Klutz complained of back pain and was sent to see the fire department doctor. The doctor diagnosed the pain as muscle strain and prescribed two weeks of medical leave.

Klutz has a record of such minor injuries. Over the past year he has been out for one reason or another at least six times. Most of these have been sprains and strains, the kind of injury that is painful but hard to see. Several have been the result of moving patients, and others have been for lifting and overreaching. On at least one occasion he claimed to have a sprained ankle from landing hard after sliding the pole at the fire station. Once he fell from a ladder, landing in a bush. He had a few scratches, but his chief complaint on that occasion was, once again, muscle pain. In this case it was likely from both the instinctive process of trying to protect himself during the fall, and from the impact of the fall.

Klutz's teammates have complained about his chronic absences. Even though another firefighter is assigned to cover his position when he is out, his teammates usually wind up doing Klutz's housework and administrative tasks. They haven't said so in so many words, but it is clear that they think he is taking advantage of the situation, and getting time off and light duty at the expense of others. Due to his absences, Klutz has missed several significant training activities and pre-incident planning visits to industrial facilities in your first-due area. As a result, he may not be as well trained as his peers.

When Klutz is around, he does good work, but is sometimes a bit overaggressive, occasionally taking action without using proper protective equipment or proper position. You have on several occasions had to remind him to don proper eye protection while using power tools.

As a good proactive leader, you have decided to tackle this problem. Klutz is a solid worker but he is causing a morale problem as well as a readiness problem for your team. Klutz has many positive qualities and you do not want to break his spirit. But you tend to agree with your teammates that he is taking advantage of the situation.

As you start the day, you are catching up on e-mail and paperwork, but Klutz's problems tend to fill your mind. What to do . . . .

What would you do?

Your assignment: Prepare a memo to your instructor outlining the steps you would take in this situation to resolve a safety problem in your company.

EMPLOYEE NOTICE OF JOB-RELATED

INJURY/ILLNESS

NAME: _____ AGENCY: _____

ADDRESS: _____ DEPARTMENT: _____

_____ SUPERVISOR: _____

HOME PHONE: _____ OFFICE PHONE: _____

Date of Incident: _____ Date of recurrence: _____

Describe accident/exposure in detail: _____

_____

_____

_____

_____

_____

Describe injuries or occupational illness: _____

_____

_____

List parts of body injured: _____

_____

_____

List names of witnesses, if any: _____

_____

I certify that all statements are true and correct to the best of my knowledge.

_____   _____
       Date   Employee

_____   _____
   Date Received   Supervisor

Completion of this form will assist in determining compensability of your claim.

Completed report should be given to your supervisor within 24 hours of your injury.

Distribution:
   Original: Claims Coordinator
             Risk Management Division
   Copy: Agency

FIN57

Figure 8-4   Most organizations provide a simple form for employees to notify their employer of job-related injuries.

Employer's First Report of Accident

Industrial Commission of Virginia
1000 DMV Drive   Richmond VA 23220
See instructions on the reverse of this form

The boxes to the right are for the use of the insurer	I.C. file number	Reason for filing
	Insurer code	Insurer location
	Insurer claim number	

Employer

1. Name of employer	City of Virginia Beach	2. Federal Tax Identification Number 540722061	3. Employer's Case No. (if applicable)
4. Mailing address	Risk Management Division Municipal Center Virginia Beach, VA 23456-9081	5. Location (if different from mailing address)	
6. Parent corporation (if applicable)		7. Nature of business Municipality	
8. Insurer Self-insured		9. Policy number	10. Effective date

Time and Place of Accident

11. City or county where accident occurred		Did accident occur on	12. Employer's premises? ☐ Yes ☐ No	13. State property? ☐ Yes ☐ No
14. Date of injury	15. Hour of injury	16. Date of incapacity	17. Hour of incapacity
18. Was employee paid in full for day of injury? ☐ Yes ☐ No		19. Was employee paid in full for day incapacity began? ☐ Yes ☐ No	
20. Date injury or illness reported	21. Person to whom reported	22. Name of other witness	23. If fatal, give date of death

Employee

24. Name of employee		25. Phone number	26. Sex ☐ Male ☐ Female
27. Address		28. Date of birth	29. Marital status ☐ Single ☐ Divorced
		30. Social security number	☐ Married ☐ Widowed
31. Occupation at time of injury or illness		32. Department	33. Number of dependent children
34. How long in current job?	35. How long with current employer?	36. Was employee paid on a piece work or hourly basis? ☐ Piece work ☐ Hourly	
37. Hours worked per day	38. Days worked per week	39. Value of perquisites per week Food/meals Lodging Tips Other	
40. Wages per hour $	41. Earnings per week (inc. overtime) $	$ $ $ $	

Nature and Cause of Accident

| 42. Machine, tool, or object causing injury or illness | 43. Specify part of machine, etc. | Were safeguards or safety equipment | 44. Provided? ☐ Yes ☐ No |
| 46. Describe fully how injury or illness occurred | | | 45. Utilized? ☐ Yes ☐ No |

47. Describe nature of injury or illness, including parts of body affected

48. Physician (name and address)	49. Hospital (name and address)	
50. Probable length of disability	51. Has employee returned to work? ☐ Yes ☐ No	If yes	52. At what wage? $	53. On what date?
54. EMPLOYER: prepared by (include name and title)	55. Date	56. Phone number
57. INSURER: processed by	58. Date	59. Phone number

This report is required by the Virginia Workers' Compensation Act

FORM NO. DF 19 REV 2/91

First Report of Accident
IC Form No. 3 (rev. 10/1/90)

Figure 8-5   Most organizations have a form for the supervisor to complete whenever an employee is injured on the job.

CHAPTER QUIZ

1. As a company officer, you can reinforce good safety practices by:
   a. bringing up past mistakes
   b. setting a good example
   c. testing your personnel
   d. transferring unsafe personnel

2. A major influence of the response safety is the _____ of the officer and the driver.
   a. attitude
   b. experience
   c. knowledge
   d. skill

3. NFPA 1500 requires that drivers of emergency vehicles, while responding to emergency situations, stop at:
   a. red lights and stop signs
   b. blind intersections
   c. railroad crossings
   d. all of the above

4. As a company officer you have a _____ responsibility for the safety and survival of your firefighters.
   a. legal
   b. limited
   c. moral and legal
   d. professional

5. The tracking of personnel as to their location and activity at an emergency scene is known as:
   a. company officer
   b. incident command
   c. personnel accountability
   d. rapid intervention

6. Tools for dealing with critical incident stress include all of the following except:
   a. counseling
   b. peer support
   c. training
   d. vacation time

7. The greatest number of firefighter injuries involve:
   a. burns
   b. smoke and gas inhalation
   c. sprains and strains
   d. wounds

8. Firefighters can make a personal commitment to their own safety by:
   a. leading a proper lifestyle
   b. maintaining good fitness and health
   c. getting proper rest
   d. all of the above

9. In order to receive the full benefit from protective equipment:

   a. it must be properly marked

   b. it must be worn and used

   c. it must be tested annually

   d. it must be discarded after five years

10. In addition to wearing full protective clothing at structural fires, it is also the responsibility of the company officer to see that all personnel wear:

   a. clean underwear

   b. leather helmets

   c. reflective stripes

   d. SCBA's

THE COMPANY OFFICER'S ROLE IN FIRE PREVENTION

CHAPTER 9

OUTLINE

- Objectives
- Overview
- What Is the Purpose of a Fire Prevention Code?
- Key Terms
- Review Questions
- Discussion Questions
- Homework
- Chapter Quiz

OBJECTIVES

Upon completion of this chapter, you should be able to describe:
- The common causes of fire and fire growth
- The requirements for fire prevention and building codes
- The fire prevention ordinances applicable to your jurisdiction
- How to deal with inquiries from the public on fire prevention issues
- How to conduct a fire prevention inspection

OVERVIEW

What can fire prevention accomplish and what is the company officer's role in fire prevention? Hopefully by now, you have a better idea of the nature of this nation's fire problem and what we can do to reduce the loss. We have seen that public fire and life-safety education programs, plans, reviews, and code-enforcement inspections all have a place in a comprehensive fire prevention program. When properly done, fire loss will be reduced. There will be fewer fires, and those that do occur will be less severe. As a result, fewer civilians will die and be injured. Equally important, fewer firefighters will die and be injured. What greater motivation could there be?

In the concluding chapter of his book *Introduction to Fire Prevention,* now in its sixth edition, author James C. Robertson challenges students to think about the problems associated with proving the value of fire prevention. The chapter, appropriately entitled "Providing Fire Prevention Works," notes that fire prevention programs cost money and, all too often, fire officials fail to recognize its value or defend its need during tough budget times. Indeed, it is a challenge. How does one track the value of life-safety education programs when it may be years between the time the safety message is delivered and the opportunity occurs for someone to use the information?

Professor Robertson suggests that most fires can be prevented. He points out that we have clear evidence of this in the relatively low fire-loss statistics for nearly every occupancy class except residential. Fires in residential occupancies remain this country's biggest fire challenge, and even here we are making progress.

111

Figure 9-1 Public education programs can be an effective way of talking to the community about fires resulting from human behavior.

Let us conclude on that note. We may fail to appreciate the value of fire prevention, and at times may wonder if we are really accomplishing any good. When we step back just a little and examine what has occurred in recent times, we will see that fire prevention does work, and that we are indeed making progress.

Every fire officer and firefighter in the fire service has an opportunity to make a significant impact on the fire-loss situation in the community. Fire prevention can save lives and property just as much as an aggressive interior attack or a dramatic rescue. Your actions may not appear on the front page of a newspaper, but they will make a difference to the community you serve (see **Figure 9-1**).

WHAT IS THE PURPOSE OF A FIRE PREVENTION CODE?

A fire prevention code provides a reasonable level of protection for life safety and property in the event of fire, explosion, or other related emergencies. Fire codes exist to minimize the hazards to life and property due to fire and panic, exclusive of those hazards considered in the building code regulations. They regulate the storage, handling, production, and use of materials by safeguarding them from fire, and they set the norms for installing, testing, and maintaining fire suppression, fire detection, and fire warning systems (see **Figure 9-2**). Codes provide specified access for fire apparatuses, define the number and location of hydrants, and so on.

Codes usually authorize the fire department to assume responsibility for the code enforcement and code administration duties within the jurisdiction served by the department. Codes are adopted by state or local government. For example, the Code of Virginia directs the Board of Housing and Community Development (an executive agency of the state) to adopt and promulgate a Uniform Statewide Building Code (USBC) so as to provide mandatory, statewide, and uniform regulations for the construction, maintenance, and use of buildings and structures.

To satisfy this mandate, the Board of Housing and Community Development publishes two volumes of USBC. Volume I regulates new construction or when a building or structure is altered, enlarged, repaired, or converted to another use. Enforcement of Volume I is mandatory statewide. Volume II regulates existing buildings and structures by requiring that they be properly maintained to protect the occupancy from the health and safety hazards that might arise from improper maintenance or use of the building. Enforcement of Volume II is optional.

The purpose of the USBC is to ensure safety to life and property from all hazards incident to the design, construction, use, repair, removal, or demolition of buildings. However, in an effort to maintain a balance of reasonableness, the USBC states that "buildings shall be constructed at the least possible cost consistent with nationally recognized standards for health, safety, energy conservation, light and ventilation, fire safety, structural strength, and accessibility for the physically handicapped and aged."

The USBC supersedes the building codes and regulations of county municipalities as well as other political subdivisions and state agencies related to the construction, reconstruction, alterations conversion, repair, or use of building systems and the installation of equipment therein. The USBC *does not* supersede zoning ordinances

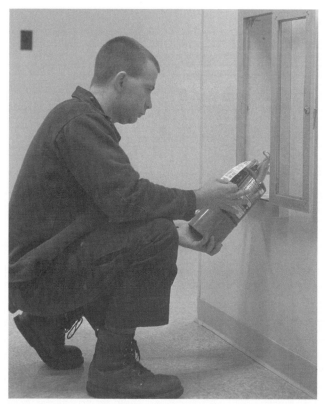

Figure 9-2 Fire suppression personnel should always check installed fire protection equipment.

or other land use controls that do not affect the manner of construction or materials to be used in the construction, alteration, or repair of buildings.

The USBC also adopts the International Building Code, Plumbing Code, Mechanical Code, and the NFPA National Electrical Code.

The Code of Virginia directs the board to adopt a mandatory Statewide Fire Prevention Code (SFPC). This code provides statewide standards safeguarding life and property from the hazards of fire or explosion arising from the improper maintenance of life safety and fire prevention and protection materials, devices, systems and structures, and the unsafe storage, handling, and use of substances, materials and devices including explosives and blasting agents. The regulations in the SFPC provide for the protection of life and property from the hazards of fire and explosion; for the safe handling, storage, and use of explosives and blasting agents; for the administration and enforcement of such regulations; and for the adoption of the International Fire Prevention Code.

Local governments are authorized to adopt fire prevention regulations that are more restrictive or more extensive in scope than the SFPC, provided such regulations do not affect the manner of construction or the materials to be used in the erection, alteration, repair, or use of a building or structure. Any provision of the SFPC which is in conflict with the USBC or other applicable law of the Commonwealth of Virginia is invalid.

The SFPC authorizes the appointment of an officer called the *local fire official* who is authorized to conduct an investigation of the origin and cause of every fire occurring within the limits of his or her jurisdiction. The state fire marshal and his assistants are authorized to enforce the SFPC in jurisdictions where local governments do not enforce the code.

While the process described here is obviously unique to Virginia, most states have a similar process. A state agency is responsible for the code, and they in turn adopt a model code. A majority of the states have statewide codes for construction, fire prevention, and the traditional trades—electrical, mechanical, and plumbing.

The chart in **Figure 9-3** details the latest adoptions at the time of printing.
Key to the more commonly adopted codes

IBC   International Building Code
IRC   International Residential Code

International Codes - Adoption by State

ICC makes every effort to provide current, accurate code adoption information, but in some cases jurisdictions do not notify ICC of adoptions, amendments or changes to their codes. To ensure you have accurate information, please contact the jurisdiction directly.

X = Effective Statewide   A = Adopted, but may not yet be effective   L = Adopted by Local Governments

State	IBC	IRC	IFC	IMC	IPC	IPSDC	IFGC	IECC	IPMC	IEBC	ICCPC	IUWIC	IZC	ICCEC	Comments
Alabama	L	L	L	X	L		L	L	L		L		L	L	
Alaska	X	L	X	X	L		L								
Arizona	X*	L	L	L	L		L	X	L					L	* State Department Health has adopted for Hospitals
Arkansas	X	X	X	A*											*Effective October 2003
California															
Colorado	L*	L	L*	L	L	L	L	L	L				L	L	* Colorado Division of Fire Safety
Connecticut				X	X										
Delaware	L	L		L	X		L	L	L						
D.C.	X	X	X	X	X		X	X	X						Effective Jan. 9, 2004
Florida				X	X		X								
Georgia	X	X	X	X	X		X	X							
Hawaii															
Idaho	X	X	X	X				X							
Illinois	L	L	L	L	L	L	L	L	L	L			L	L	
Indiana	X	X	X	X			X								
Iowa	L	L	L	L	L	L	L	L	L				L	L	
Kansas	X*	L	L	L	L	L	L	X	L						* IBC acceptable for state except for schools
Kentucky	X	X		X				L							
Louisiana	X			X											
Maine	L	L	L	L	L	L	L	L	L				L	L	
Maryland	X	X		L	L		L	L							
Massachusetts	A	A		A											
Michigan	X	X	X	X	X		X		X	X					
Minnesota	X	X	X												
Mississippi	L	L	L	L	L	L	L	L	L						
Missouri	L	L	L	X*	X*	L	L	L	L		*			L	*State buildings only
Montana	X	X													
Nebraska	L	L	L	L	L	L	L	L*	L	L			L	L	* State owned or funded buildings Effective January 1, 2004
Nevada	L	L	L	L	L		L								
New Hampshire	X	L	L	X	X		L	X							
New Jersey	X	X		X			X								
New Mexico	L	L	L	L	L		L		L						
New York	X	X	X	X	X		X	X	X						
North Carolina	X	X	X	X	X		X	X							
North Dakota	X	X	L	X			X		L						
Ohio	X	L	L	X	X		X		L						
Oklahoma	X	X*	X	X	X	L	X	L	X	X			L	L	* Mechanical provisions only
Oregon		X	A*	X			X								*Effective 10-1-04
Pennsylvania	A	A	A	A	A	L	A	A		A		A	L	A	Effective 4/09/2004
Rhode Island	X*	X		X	X		X	X							*IBC used for Rehab Code
South Carolina	X	X	X	X	X	X	X	X	X	X	X	X	X	X	Effective 01-01-04/2003 editions
South Dakota	X*	L	X*	L	X		L	L	L	L					*Approved for local adoption
Tennessee	L	L	L	L	X		L	L	L	L					
Texas	L**	X**	L	X*	X*	L	L	X*	L				L	L	*IMC & IPC approved for local adoption ** TX Dept. of Insurance
Utah	X	X	X	X	X		X	X							2003 Editions
Vermont				X											
Virginia	X	X	X	X	X		X	X	X			L		X	
Washington	X	X	X	X			L			L		L			
West Virginia	X	X		X	X		X	X	X	X					
Wisconsin	X		L	X			X	X							
Wyoming	X	L	X	X	L		X		L				L		
Dept. of Defense	X														Unified Facilities Criteria
Nat'l. Park Service	X														
Puerto Rico					X										

Figure 9-3 International Codes—adoption by state.

IFC   International Fire Code

IMC   International Mechanical Code

IPC   International Plumbing Code

IPSDC   International Public Sewage Disposal Code

IFGC   International Fuel Gas Code

IECC   International Energy Conservation Code

IPMC   International Property Maintenance Code

Key Terms

Define in your own words the following terms:

arson _____

_____

building code _____

_____

certificates _____

_____

Dillon's rule _____

_____

fire prevention _____

_____

fire prevention code _____

_____

fire suppression _____

_____

in-service company inspection _____

_____

licenses _____

_____

life-safety education _____

_____

minimum standards _____

_____

permits _____

_____

residential _____

_____

specifications _____

_____

standard of performance _____

_____

QUESTIONS

Review Questions

1. What are the two ways a fire department can protect its community from fire loss?

_____

_____

_____

_____

_____

2. The Commission on Fire Prevention and Control identified three problems that inhibit fire prevention activities. What were they?

_____

_____

_____

_____

3. Name four ways fire loss is reported. What are recent values for each of these?

_____

_____

_____

_____

4. What are the leading causes of fire?

_____

_____

_____

_____

5. What is the leading cause of fire deaths?

_____

_____

_____

6. Where do most of these deaths occur?

_____

_____

_____

_____

7. How has the smoke detector reduced this statistic?

_____

_____

_____

_____

8. What are the three "E's" of fire prevention?

_____

_____

_____

_____

9. How are fire prevention codes adopted?

_____

_____

_____

_____

10. How can you, as a company officer, boost fire prevention efforts in the community you serve?

_____

_____

_____

_____

Discussion Questions

1. What are the most common causes of fire in your community?

_____

_____

_____

2. What are the fire prevention programs in your community? How effective are they?

_____
_____
_____
_____

3. How does your community's fire loss compare with the national statistics?

_____
_____
_____
_____

4. Identify five social or economic factors that influence the fire problem in a typical community. How do they impact on fire loss?

_____
_____
_____
_____

5. What role has the smoke detector had in reducing fire loss? What problems remain regarding smoke detectors and fire loss?

_____
_____
_____
_____

6. In addition to the direct dollar loss associated with fires, what other types of economic losses occur?

_____
_____
_____
_____

7. Why does the United States have such high fire-loss statistics?

_____
_____
_____
_____

8. What codes (building and fire prevention are the most relevant) are in effect in your state?

_____

_____

_____

_____

9. In your state, can local jurisdictions change the provisions of the building code or the fire prevention code?

_____

_____

_____

_____

10. Explain the code adoption process in your state.

_____

_____

_____

_____

HOMEWORK FOR CHAPTER 9

For Fire Officer I

Deliver a Public Education Program

What NFPA 1021 says:

Section 4.3. "This duty involves dealing with inquiries of the community and projecting the role of the department to the public and delivering safety, injury, and fire prevention education programs, according to the following job performance requirements."

Specifically, section 4.3.4 says that the Fire Officer I shall "Deliver a public education program, given the target audience and topic, so that the intended message is conveyed clearly."

Situation:

You have been assigned to represent your fire department at a meeting of the local Junior Chamber of Commerce. The club gathers each Tuesday for a dinner meeting. Because of a recent fatal fire in the community, the club has indicated an interest in the nature of the fire problem at the national, state, and local level. You are to speak for 10 to 15 minutes.

As you analyze your assignment, you decide that your task is to prepare for and present a lesson, and will ideally be able to use visual aids to present some statistical data. You realize that during such a presentation, you should be able to understand and clearly summarize relevant current local, state, and national statistics. Sources of up-to-date information include data published by the NFPA and the NFIRS as well as state and local sources. Local data may be the most challenging to obtain, but is the most meaningful because it focuses on the local fire problem, and may be of greatest interest. It would also be beneficial to learn as much as you can about your audience before the event.

Your assignment: Prepare a lesson plan to use during your presentation.

Suggested topics to include are as follows:

The common causes of fire

Basic safeguards for fire prevention in the home

Examples of common events and how they could have been prevented

Illustrative stories of recent fires in the community

Basic instructions on what to do if fire does occur

Your instructor may ask you to actually present some or all of your material to the class.

Note: In some jurisdictions, evidence of certification as a Fire and Life Safety Educator I (or higher) in accordance with NFPA 1035 may be substituted for this requirement.

For Fire Officer II

Conduct a Fire-Code Inspection

What NFPA 1021 says:

Section 5.5 "This duty involves conducting inspections to identify hazards and address violations and conducting fire investigations to determine origin and preliminary cause, according to the following job performance requirements." (Note—there are two separate activities here.)

Specifically, section 5.5.1. says that the Fire Officer II shall "describe the procedures for conducting fire inspections, given any of the following occupancies, so that all hazards, including hazardous materials, are identified, approved forms are completed, and approved action is initiated:

1. Assembly
2. Educational
3. Health care

4. Detention and correctional
5. Residential
6. Mercantile
7. Business
8. Industrial
9. Storage
10. Unusual structures
11. Mixed occupancies

Situation:
The level of inspection activity done by suppression personnel varies widely. You may or may not have had experience in conducting inspections, but to meet this requirement, you should conduct an inspection of a real building. You can certainly work with others on this project, but you should complete and submit your own inspection report.

Your instructor may provide additional direction.

Your assignment: Perform an inspection of an occupancy in your area and submit the completed inspection report to your instructor as evidence of your accomplishment of this activity.

If your department does not routinely perform inspections, you may perform a courtesy inspection on a business occupancy in your area. If you are unable to perform an inspection of a business occupancy, perform an inspection on a fire station, public building, or church. Submit a copy of the completed inspection report to your instructor as evidence of your accomplishment of this activity.

Determine the Point of Origin and Preliminary Cause of Fire

Specifically section 5.5.2 of NFPA 1021 says that the Fire Officer II shall "Determine the point of origin and preliminary cause of a fire, given a fire scene, photographs, diagrams, pertinent data and/or sketches, to determine if arson is suspected."

This requirement is addressed in Chapter 11.

CHAPTER QUIZ

1. Efforts designed to prevent fire from occurring, or to minimize the loss when fire does occur, are called:
   a. fire control
   b. fire inspection
   c. fire prevention
   d. fire suppression
2. Most fires occur:
   a. in residences
   b. in structures
   c. in vehicles
   d. outdoors
3. Most fire fatalities occur:
   a. in residences
   b. in structures
   c. in vehicles
   d. outdoors

4. The leading cause of fire in residence is associated with:
   a. arson
   b. cooking
   c. heating
   d. smoking

5. The three Es of fire prevention, representing the three best-known components of a well-established fire prevention program, include all of the following except:
   a. education
   b. enforcement
   c. engagement
   d. engineering

6. Model fire codes are written:
   a. as a federal law
   b. as a local law
   c. as a minimum standard
   d. for specific fire departments

7. In order to be enforceable, the fire prevention code must be:
   a. adopted into law
   b. approved by local government
   c. approved by the fire department
   d. published in the newspaper

8. A fire prevention code:
   a. is a legal document
   b. regulates the storage and use of hazardous materials
   c. sets forth minimum requirements for fire protection
   d. all of the above

9. The first objective of fire prevention is:
   a. confining fire to a limited area
   b. preventing property damage
   c. reducing insurance rates
   d. safeguarding life against fire

10. In order to develop an effective local public fire education program the first step should involve:
   a. an analysis of national statistics
   b. defining the local fire problem
   c. developing clear objectives
   d. soliciting citizens to gain financial support

THE COMPANY OFFICER'S ROLE IN UNDERSTANDING BUILDING CONSTRUCTION AND FIRE BEHAVIOR

CHAPTER 10

OUTLINE

- Objectives
- Overview
- What Makes the Fire Go Out?
- Key Terms
- Review Questions
- Discussion Questions
- Homework
- Chapter Quiz

OBJECTIVES

Upon completion of this chapter, you should be able to describe:
- Basic types of building construction used by the fire service
- The strengths and weaknesses of each building type
- Basic fuel loading in structures
- Fire ignition, growth, and development in structures
- Fire control and extinguishment

OVERVIEW

In Chapter 8 we discussed firefighter safety. Many of the accidents that injure firefighters occur at the fireground, often during suppression operations. In Chapter 9 we discussed fire prevention and recognized that there are opportunities for reducing the number of fires and the severity of fires that do occur. In theory, if our prevention activities were completely successful, we would have few, if any, fires. But even with the best of prevention programs, fires still occur. While most fires occur outside, most fire departments focus their resources on structural firefighting. A good understanding of building construction and fire behavior is essential.

If you like to read about the history of firefighting, you have likely seen pictures of fire suppression activity showing a wall collapse. These were very dramatic events and they took the lives of many firefighters. Today, we are more likely to see a floor collapse. These can also be dramatic, especially for personnel operating inside at the time of the event. Modern lightweight building construction not only makes such collapses more likely, it makes it more likely to occur earlier in the fire.

To keep from getting caught in a collapse situation, firefighters should understand the type of construction used in buildings in their response area, as well as the way fire affects these buildings. They should also understand

Figure 10-1 Knowing your building is important. Looking at the front of this building, especially at night, might create the wrong impression about its roof structure.

how likely fire is to travel through the building and choose tactics that are appropriate for the situation at hand in order to stop the fire's destructive advance before the collapse occurs.

Many firefighter fatality investigations note that firefighters need more training on building construction and fire behavior. We look to you as fire officers to provide that training so that your personnel are operating safely and effectively at structural fires.

At the scene of emergencies, the first arriving officers are usually concerned with limited resources, time, and many unknown factors. Working in these conditions is difficult and dangerous; leading others under such conditions can be extremely challenging. In such adverse conditions, the company officer is expected to be calm and decisive, to issue clear orders, and to keep track of all the activities. The company officer is also responsible for the safety of others, some aspects of incident management, and in many situations, may be the incident manager.

To be effective in these roles, the company officer must have a thorough understanding of fire behavior and building construction (see **Figure 10-1**). These areas of knowledge are fundamental to the task of fire suppression. Understanding and anticipating the forces that control the growth and extension of a fire will also facilitate a more effective and safer extinguishment operation. Company officers should keep these items in mind as they plan, train, and operate with their companies at burning buildings.

WHAT MAKES THE FIRE GO OUT?

Most structure fires are extinguished by using water to absorb the heat created by the fire (see **Figure 10-2**). From basic training, we recall that reducing the temperatures of items burning below their ignition temperatures will cause the fire to go out. To extinguish a fire, the quantity of water must exceed the heat being produced. If we can estimate the amount of heat being produced, we can estimate the amount of water needed. Some fire departments apply water in ever-increasing rates until they reach a point where the water starts to affect the fire. This trial-and-error approach takes time during which the fire is destroying property. During preplanning and size-up, you have an opportunity to estimate the needed water flow in advance. This estimate is theoretical, of course, but it is a better starting point than nothing at all. The amount, or rate, of water needed is called the *theoretical fire flow*.

Figure 10-2 We commonly use water to extinguish fires. The water flow must be sufficient to overcome the heat produced by the burning materials.

$$\text{The basic formula: Fire flow (in gpm)} = \frac{\text{involved area}}{3}$$

Take, for example, a fully involved, one-story building that is 30 by 40 feet. Using the formula above, we multiply 30 times 40 and divide by 3, or 400. Thus, the theoretical fire flow is 400 gpm.

What can we do with this information? Right away, you might be able to answer the following important questions:

▍ What hoseline will I use?

▍ Do I have enough personnel to safely deploy the line?

▍ Do I have sufficient water to sustain the flow until the fire is under control?

▍ Should I ask for a second alarm?

This simple formula will help answer all of these questions.

Consider what might be needed for additional floors. Assuming we have a 30 by 40 foot building with *two* fully involved floors, we will need to double our calculations and require twice as much water, or 800 gpm.

Hopefully, the fire has not extended to a "fully involved" status upon your arrival. Let us say you have a 30 by 40 foot building with the fire fully involved in one room and extending to the floor above. In this case, you might be able to estimate that each floor is 25 percent involved. Hence, we can reduce our water flow accordingly.

$$\text{25 percent of } \frac{30 \times 40}{3} = 100 \text{ gpm for each floor, or 200 gpm}$$

This calculation assumes two important points. First, it assumes that there is a fire-separation wall between the fire and the rest of the building in horizontal and vertical directions. It also assumes that your fire attack is effective in confining the fire to the area already burning.

Exposures

If we have exposures, we will need to protect them, and this will require additional water flow. We may have exposures in terms of areas that are threatened but not yet involved *within the fire building* and in *adjacent buildings.* Let us assume you have a well involved "room and contents" that requires 100 gpm. Because the door of the room is open, the fire is rapidly advancing to the adjacent room. Assume the room is an exposure (at least) and needs to be protected. We may have the same situation in the room above the fire, especially in residential construction.

For protecting adjacent buildings, we can decide how much of a risk is presented. If a structure is within 30 feet of the fire building, we will need to consider the exposure as threatened and provide some protection. Under normal conditions, only one side of the exposure building is presented to the fire, so we need to protect only one side, or 25 percent. We can determine the fire flow needs of the exposures by multiplying the fire flow of the involved building by 25 percent for each exposure.

If the exposure building is more than 100 feet away, it is unlikely to be seriously threatened. For those buildings between 30 and 100 feet away, some judgment will be required. Factors to consider may include the amount of fire involvement, the building material used on the exterior of the exposure building, the number and size of openings in the exposure building, the direction of the wind, and so on. Realize that if the exposure becomes involved, we will have another fire building and we will need separate calculations and more water for that building.

Example 1. Suppose your fire is in a townhouse and that the unit involved has exposures on both sides. You determine that the needed fire flow is 600 gpm for the fire building. According to these calculations, you should also consider that each of the adjacent units presents you with an exposure situation, and each would require 150 gpm of fire flow.

Thus, your total will be 600 + 150 + 150, or 900 gpm.

Example 2. Let us examine a 30 by 50 foot commercial building. The basic formula would be 30 by 50, divided by 3, or 500 gpm. If there are two stories, and both are fully involved, we will need twice that much, or 1,000 gpm. Suppose we have two similar detached buildings that are located on either side of our fire building. We will need water to protect each of them, and we can estimate 25 percent of the water will be needed for the fire building, or 250 gpm. We have 1,000 + 250 + 250, or 1,500 gpm.

Let us use our heads here for a minute. What does the fire flow tell the officer on the first arriving pumper? For openers, it will tell you what hoseline to pull. It is unlikely that you will do much good with a 1¾-inch hoseline in either case. It would be better to start with something bigger because we need to sustain the flow for several minutes, at least. If you arrive with a 500- or 750-gallon tank on your pumper, you are going to run out of water in a couple of minutes. Get a reliable water supply set up right away.

Personnel Requirements

These numbers help us determine the size of the hoselines we will use and help us estimate our water supply requirements. In addition, these numbers will help us determine our personnel needs. If we assume that each 1½- or 1¾-inch hoseline will require at least two firefighters, and that each 2½-inch hoseline will require three or four firefighters, we can quickly determine how many firefighters are needed for fire extinguishment. Additional firefighters will be needed for search and rescue, ventilation activities, water supply and pumping, command, and so on.

Going back to our townhouse example, we need 600 gpm in the fire building and 150 for each exposure. Let us assume that we are getting 150 gpm from each 1¾-inch hoseline we place in service. Initially, we will need at least four 1¾-inch lines in the fire building and one in each exposure to provide the desired fire flow. Six 1¾-inch lines will require twelve firefighters. Using two 2½-inch hoselines for fire attack and two 1¾-inch hoselines for exposure control would be more appropriate in many departments, but you will still need ten firefighters to advance the hoselines. We should note that these numbers do not leave any room for error, nor do they provide any backup hoselines for safety.

Realize that we have spent more time discussing this than the time you have to make important decisions during a fire. The basic formula provides a tool to help make quick decisions that might otherwise be based on instinct. You can practice the mental part of the concept and apply it in your preplanning.

Let us go back to the four big size-up questions and apply them to our commercial situation.

> **What hoseline will I use?** With 500 gpm needed on each of two floors, you should consider several 2½-inch hoselines. Using a similar number of smaller hoselines will not provide the needed fire flow.

> **Do I have enough personnel to safely deploy the line?** We do not know how many you brought, but you will need at least three people on each of the 2½-inch hoselines and two for each 1¾-inch hose-

line. Now add a few more people for other things like ventilation, water supply, and so on. You do the math.

Do I have a sufficient water supply to sustain that flow until the fire is under control? Flowing 1,500 gpm is easy for many urban departments, but will be a serious challenge for many others. You will need a reliable water supply and effective pumping capacity to sustain the attack.

Should I ask for a second alarm? What do you think?

Key Terms

Define in your own words the following terms:

access factors _____

_____

automatic fire protection sprinkler _____

_____

backdraft explosion _____

_____

BTU _____

_____

community consequences _____

_____

fire behavior _____

_____

fire extension _____

_____

fire load _____

_____

fire resistive _____

_____

flashover _____

_____

free-burning phase _____

_____

fuel load _____

_____

heat of combustion _____

_____

heavy timber construction _____

_____

incipient phase _____

_____

latent heat of vaporization _____

_____

life risk factors _____

_____

noncombustible construction _____

_____

occupancy factors _____

_____

ordinary construction _____

_____

physical factors _____

_____

piloted ignition temperature _____

_____

property risk factors _____

_____

resource factors _____

_____

rollover _____

_____

size-up _____

_____

smoldering phase _____

_____

specific heat _____

_____

standpipe systems _____

_____

structural factors _____

_____

survival factors _____

_____

theoretical fire flow _____

_____

thermal stratification _____

_____

ventilation _____

_____

woodframe construction _____

_____

QUESTIONS

Review Questions

1. What is meant by *life-risk* factors?

_____

_____

_____

_____

2. What is meant by *property risk* factors?

_____

_____

_____

_____

3. What is meant by *community consequences*?

_____

_____

_____

_____

4. How do each of the following factors affect firefighting operations?

Physical factors _____

Access factors _____

Structural factors _____

Survival factors _____

5. Define the five building classifications.

_____

_____

_____

_____

6. Identify a fire-related hazard associated with each type of building.

_____

_____

_____

_____

7. Define the following:

Rollover _____

Flashover _____

Backdraft _____

8. What is the National Fire Academy's fire flow formula?

_____

_____

_____

_____

9. How does the NFA fire flow formula help in prefire planning and fire suppression operations?

_____

_____

_____

_____

10. Why is an understanding of building construction and fire behavior essential for you, as the company officer?

_____

_____

_____

_____

Discussion Questions

1. What are the significant risk factors in your community?

_____

_____

_____

_____

2. What are the five principal building types? What is the significant problem from a firefighter safety perspective while firefighting in each type?

_____

_____

_____

_____

3. Which types of buildings are found in your community? What are the significant hazards of these buildings?

_____

_____

_____

_____

4. List some of the target hazards in your community.

_____

_____

_____

_____

5. How has building construction changed in your community in the last 50 years?

_____

_____

_____

_____

6. How have the contents of buildings changed in the last 50 years?

_____

_____

_____

_____

7. How have these changes affected fire suppression activities?

_____

_____

_____

_____

8. We hear a lot about firefighters going where they have never gone before—that firefighters, protected by better protective clothing and SCBA, can penetrate further into a building to locate and control fires. In light of your answer to the previous question, how has firefighting changed in the last 10 years?

_____

_____

_____

_____

9. Explain the use of the fire flow formula. What formula is used in your community?

_____

_____

_____

_____

10. Have you used a fire flow formula during preplanning activities? Have you used it during size-up and fire suppression activities?

_____

_____

_____

_____

HOMEWORK FOR CHAPTER 10
Maintaining Basic Skills

For Fire Officer I and II

This chapter reviewed basic knowledge regarding building construction and fire behavior. Let us continue to review basic knowledge and skills by looking at various basic evolutions expected of firefighters and companies in any good fire department.

Prepare a short outline (essentially a lesson plan) of the information needed by your personnel to perform that event. Your outline should identify the purpose of the activity, the equipment needed, and the process by which success will be measured.

Submit your outline to your instructor for evaluation.

There is a much more comprehensive set of basic evolutions that can be used for training and evaluating a company's performance. Namely, NFPA 1410, *Standard on Training for Initial Emergency Scene Operation*. NFPA 1410 is designed to provide fire departments with an objective method of measuring performance for initial fire suppression and rescue procedures.

If your department is using NFPA 1410, you may substitute one of the events listed there for the purpose of fulfilling the requirements of this assignment.

CHAPTER QUIZ

1. The main reason for firefighters to study building construction is:
   a. to provide a second job in the construction industry
   b. to gain respect and appreciate engineering practices
   c. to understand how buildings react during fire conditions
   d. to know how to ventilate the building when needed

2. Construction using fire-resistive material, and structural components protected from fire, is known as _____ construction.
   a. fire resistive
   b. heavy timber
   c. noncombustible
   d. ordinary

3. Construction in which the walls are made of noncombustible material and the roof is supported on unprotected steel beams is known as _____ construction.
   a. fire resistive
   b. heavy timber
   c. noncombustible
   d. ordinary

4. Construction consisting of wood joists and masonry load-bearing walls is known as _____ construction.
   a. fire resistive
   b. heavy timber
   c. noncombustible
   d. ordinary

5. Construction consisting of noncombustible walls with the floor and floor rafters made of large timbers is known as _____ construction.
   a. fire resistive
   b. heavy timber
   c. noncombustible
   d. ordinary

6. The expected maximum amount of combustible material in a given fire area is called:
   a. combustible load
   b. fire load
   c. fuel load
   d. smoke load

7. Installed fire protection equipment might include:
   a. smoke detectors
   b. sprinklers
   c. standpipes
   d. all of the above

8. A thermosensitive device that releases a spray of water over an area to control, or extinguish, a fire is called a:
   a. fire alarm
   b. fire extinguisher
   c. sprinkler
   d. standpipe

9. The field of science associated with fire-related phenomena is called:
   a. fire behavior
   b. fire extension
   c. fire phenomena
   d. fire technology

10. When fire burns it consumes the oxygen in an enclosed space. When fresh air is introduced into the area, we might experience a:
   a. backdraft
   b. flashover
   c. rollover
   d. rekindle

THE COMPANY OFFICER'S ROLE IN FIRE INVESTIGATION

OUTLINE

- Objectives
- Overview
- First Responders' Opportunities for Making Observations

- Important Questions Regarding Fire Investigations
- Guidelines for Calling the Fire Investigator
- Key Terms

- Review Questions
- Discussion Questions
- Homework
- Chapter Quiz

OBJECTIVES

Upon completion of this chapter, you should be able to describe:

- The common causes of fire
- The significance of arson as part of this nation's fire problem
- How to perform a preliminary fire investigation
- How to recognize evidence that would suggest that a fire was deliberately set
- How to secure the incident scene and preserve evidence

OVERVIEW

There are several benefits of investigating fires, and all fires should be investigated. Fire investigation is a part of any prevention program and it can help identify the community's fire risks. This information can be used to develop fire prevention programs that address the community's particular needs.

Investigation will also help a department identify problems associated with fire suppression activities and may help identify additional resources needed. When arson is involved, a good fire investigation is a necessary first step in administering justice.

Company officers play an important role in conducting fire investigations. They are likely to be among the first emergency response personnel to arrive at the scene. For smaller events, they may be the senior department representative on scene.

Company officers should be aware of the causes of fire and the impact of arson on their community. They should also be able to determine the cause of most accidental fires and recognize the signs and symptoms of

Figure 11-1 Company officers should complete the fire report as soon as they can while the details of the event are still fresh so that the information in the report is complete and accurate.

arson. In either case, they should be able to conduct a basic investigation that will properly determine and document the origin and cause of the fire (see **Figure 11-1**). They should understand the rights of the property owner and the requirements for conducting a lawful investigation.

🔘 FIRST RESPONDERS' OPPORTUNITIES FOR MAKING OBSERVATIONS

1. During the initial notification of the event
   - Identification and location of the caller
   - The caller's tone, level of excitement, and accent
   - Background noises
2. While en route to the fire
   - Additional information that may be provided by caller
   - The weather, time of day, and so on
   - Delays due to highway construction, trains, and so on
3. Upon arrival
   - Any persons present and what they were doing
   - Any vehicles present and any that were leaving
   - Note the fire conditions: location and intensity of the fire, color of the flames, and smoke

4. Observation during size-up
   - Operation of any alarm or suppression equipment
   - Any unusual observations
   - Methods of escape of any occupants
5. During the initial stages of suppression
   - Were furnishings and inventory in place?
   - Was the fire alarm sounding?
   - Was the sprinkler system operating?
6. During overhaul activities
   - Recheck items noted above
   - Determine origin and cause
   - Determine path of fire travel

IMPORTANT QUESTIONS REGARDING FIRE INVESTIGATIONS

Who discovered the fire?

When was the fire first discovered?

When was the fire reported?

Who provided the first report of the fire?

Who extinguished the fire?

Who provided scene security?

Who has pertinent knowledge regarding the fire?

Who had a motive for setting the fire?

What happened during the fire?

What actions were taken by firefighters?

What damage occurred?

What do the witnesses know?

What evidence was found?

Where did the fire start?

Where did the fire travel?

Where were the occupants?

GUIDELINES FOR CALLING THE FIRE INVESTIGATOR

An investigator should be called in the following situations:

1. Any event that involves an obvious incendiary fire
2. Any fire that results in death or serious injury
3. Any event that produces burn injuries, especially those involving direct flame contact, fireworks, or is the result of an assault
4. Any event that involves an exploding or incendiary device
5. Any event that involves property damage beyond the room of origin
6. Any event that produces damage to government property
7. Any vehicle fire that is not the result of an accident
8. Any event in which the officer in charge is unable to determine the cause

In any event in which the fire officer has concerns about these or other issues, he or she should make an effort to talk with a fire investigator by phone or other means before leaving the scene.

Key Terms

Define in your own words the following terms:

accidental causes _____

_____

administrative process _____

_____

administrative search warrant _____

_____

arson _____

_____

energy _____

_____

motive _____

_____

oxidation _____

_____

probable cause _____

_____

search warrant _____

_____

smoking materials _____

_____

vented _____

_____

QUESTIONS

Review Questions

1. What are the most common causes of fire?

_____

_____

_____

_____

2. What is *arson*?

_____

_____

_____

_____

3. What is meant by the term *motive*?

_____

_____

_____

_____

4. What are some common motives for arson fires?

_____

_____

_____

_____

5. What information should be noted by the first arriving personnel at the scene of any fire?

_____

_____

_____

_____

6. What information should be noted by fire personnel during fire suppression operations?

_____

_____

_____

_____

7. What information should be noted during overhaul activities?

_____

_____

_____

_____

8. What are the ways you can enter private property to conduct a fire investigation?

_____

_____

_____

_____

9. Why is it important to complete a fire incident report for every event?

_____

_____

_____

_____

10. What is your role as the company officer in fire cause determination?

_____

_____

_____

_____

Discussion Questions

1. Why should firefighters be concerned about intentionally set fires?

_____

_____

_____

_____

2. What causes people to set fires?

_____

_____

_____

_____

3. What is the cost of these intentionally set fires in this country?

_____

_____

_____

_____

4. What is the extent of these intentionally set fires in your community?

_____

_____

_____

_____

5. What can be done to reduce the incidence of such fires in your community?

_____

_____

_____

_____

6. Explain how a fire should be documented.

_____

_____

_____

_____

7. What legal considerations must be kept in mind while searching for evidence of the origin and cause of a fire?

_____

_____

_____

_____

8. Discuss the impact of the _Michigan vs. Tyler_ case as it applies to the actions that should be taken by the incident Commander during and following a fire.

_____

_____

_____

_____

9. Under what circumstances would you call a fire investigator?

_____

_____

_____

_____

10. As a company officer, what can you do to reduce the number of intentionally set fires in your response area?

_____

_____

_____

_____

HOMEWORK FOR CHAPTER 11
Inspection and Investigation

For Fire Officer I
What NFPA 1021 Says:
Section 4.5

This duty involves performing a fire investigation to determine preliminary cause, securing the incident scene, and preserving evidence, according to the following job performance requirements.

Specifically, section 4.5.1 says that the Fire Officer I shall "evaluate available information, given a fire incident, observations, and interviews of first-arriving members and other individuals involved in the incident, so that a preliminary cause of the fire is determined, reports are completed, and, if required, the scene is secured and all pertinent information is turned over to an investigator.

Section 4.5.2 says that the Fire Officer I shall "Secure an incident scene, given rope or barrier tape, so that unauthorized persons can recognize the perimeters of the scene and are kept from restricted areas, and all evidence or potential evidence is protected from damage or destruction."

Situation:
At 2:34 P.M. your engine company was called to deal with a structure fire. Your company was the first to arrive. You found smoke and fire conditions in the kitchen of a second floor apartment. Thanks to the presence of an operating smoke detector with on-premises monitoring by a resident manager, the fire department was quickly notified of the situation. Upon entry you conducted an initial search of the apartment and found that no one was present. It appeared that the apartment was unoccupied—while it was furnished, there were no personal possessions save a few boxes, suggesting that someone was moving in or out.

The truck company from your station assisted you in your suppression efforts. The battalion chief responded, but because of the quickness and efficiency of your efforts, declined to take command and place the other engine company in service. This left you as the incident commander, and responsible for gathering the information needed for the fire report.

The apartment manager, Mr. Rain Hathaway, reported he had just rented the apartment to a Mrs. Sunny Day, and in fact she had picked up her key and started to move in. He did not have an address or phone number where she could be reached. As your company was taking up, a lady showed up with a van full of boxes. She and several friends made casual conversation about all the firefighters, and discovered that the scene of the firefighter's efforts was her new apartment. Mrs. Day approached you to ask about what had happened.

During your conversation with her, you learned that she and several friends had made several trips during the day, bringing in boxes they could manage. They had last been at the apartment about two hours earlier. During that trip they brought in several boxes of kitchen items and left them in the kitchen. Then they left, and everything seemed normal.

Since you carry a digital camera in the apparatus, you took photographs during overhaul activities as the area was cleared layer by layer. Later you printed these out for your crew to look at as you asked them to help you figure out what happened and write the report (See **Figure 11-2**).

Your assignment: Prepare a fire incident report for this call. The date is today's date and the weather is the current weather. The address is 108 Main Street in your city. The fire took place in apartment 21. Use the report used by your fire department or the form provided (**Figure 11-3**). Suggestion: Make a copy of the incident report, complete it, and submit a neat completed form to your instructor.

For Fire Officer II
What NFPA 1021 says: section 5.5 "This duty involves conducting inspections to identify hazards and address violations and conducting fire investigations to determine origin and preliminary cause, according to the following job performance requirements."

Specifically, section 5.5.2 The Fire Officer II shall "determine the point of origin and preliminary cause of a fire, given a fire scene, photographs, diagrams, pertinent data and/or sketches, to determine if arson is suspected."

Situation:
Same as above.

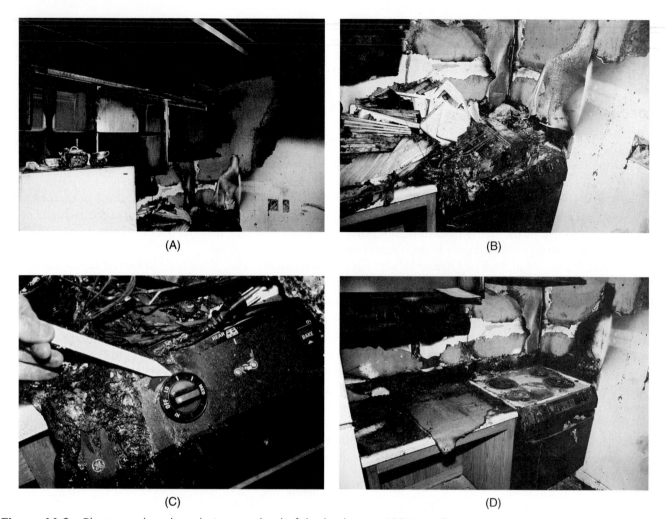

(A) (B)

(C) (D)

Figure 11-2 Photographs taken during overhaul of the kitchen at 108 Main Street.

Your assignment: Using the same information as provided for the Fire Officer I, determine the point of origin and preliminary cause of a fire, as well as if arson is suspected.

Present this information in a memo report to your instructor using the following format:

Preliminary statement. Briefly recap the scene—the who, what, when, and so on.

Findings of fact. What did you learn while putting the facts together? Who did you talk to, what did they tell you, and what did you learn from the pictures? If you have facts, put them here. Save your opinion for the next section.

Opinions. Based on the facts, express your opinions including the ultimate opinion: What do you think happened; what was the origin and cause of the fire?

Recommendations. What follow-up action is needed here? Does anything else need to be done to close this case? Is there any reason to have a fire investigator review this situation, or is there any reason to suspect that this fire was deliberately set?

North Carolina Incident Report

A — Agency ID | State | Incident Date (MM DD YYYY) | Station | Incident Number | Exposure | ☐ Delete ☐ Change ☐ No Activity | BASIC –1

B Location — ☐ Check this box to indicate that the address for this incident is provided on the Wildland Fire Module in **Section B** "Alternative Location Specification." | Census Tract

☐ Intersection
☐ Block Address
☐ In front of
☐ Rear of
☐ Adjacent to
☐ Directions

Number/Milepost | Prefix | Street or Highway | Street Type | Suffix
Apt./Suite/Room | City | State | Zip Code
Cross street or directions, as applicable

C Incident Type — Incident Type

D Aid Given or Received
1 ☐ Mutual aid received
2 ☐ Automatic aid recv.
3 ☐ Mutual aid given
4 ☐ Automatic aid given
5 ☐ Other aid given
N ☐ None
Their FDID | Their State | Their Incident Number

E1 Dates & Times — Midnight is 0000. Check boxes if dates are the same as Alarm Date.
Month Day Year Hour Min
Alarm — ALARM always required
☐ Arrival — ARRIVAL required, unless canceled or did not arrive
☐ Controlled — CONTROLLED optional, except for wildland fires
☐ Last Unit Cleared — LAST UNIT CLEARED, required except for wildland fires

E2 Shifts & Alarms — Local Option — Shift or platoon | Alarms | District

E3 Special Studies — Local Option — Special Study ID# | Special Study Code

F Actions Taken
Primary Action Taken (1)
Additional Action Taken (2)
Additional Action Taken (3)

G1 Resources — ☐ Check this box and Complete the Apparatus form
Apparatus | Personnel
Suppression | EMS | Other
☐ Check box if resource counts include mutual aid resources.

G2 Estimated Dollar Losses & Values
LOSSES: Required for all fires. Otherwise Optional | None
Property $ ☐
Contents: $ ☐
PRE-INCIDENT VALUE: Optional
Property $ ☐
Contents: $ ☐

H1 Casualties ☐ None
Deaths | Injures
Fire Service
Civilian

H2 Detector — Required for confirmed fires
☐ Detector alerted occupants

H3 Hazardous Materials Release
N ☐ None
1 ☐ **Natural gas:** slow leak, no evacuation or hazmat actions
2 ☐ **Propane gas:** <21 lb. tank (as in home BBQ grill)
3 ☐ **Gasoline:** vehicle fuel tank or portable container
4 ☐ **Kerosene:** vehicle fuel tank or portable container
5 ☐ **Diesel fuel/fuel oil:** vehicle fuel tank or portable storage
6 ☐ **Household solvents:** home/office spill, cleanup only
7 ☐ **Motor oil:** from engine or portable container
8 ☐ **Paint:** from paint cans totaling <55 gallons
0 ☐ **Other:** Special Hazmat actions required or spill > 55 gal.
☆ Please complete the Hazmat form

I Mixed Use Property
NN ☐ Not Mixed
10 ☐ Assembly use
20 ☐ Education use
33 ☐ Medical use
40 ☐ Residential use
51 ☐ Row of stores
53 ☐ Enclosed Mall
58 ☐ Business & residential
59 ☐ Office use
60 ☐ Industrial use
63 ☐ Military use
65 ☐ Farm use
00 ☐ Other mixed use

Completed Modules
☐ Fire-2 ☐ Hazmat-7
☐ Structure-3 ☐ Wildland Fire-3
☐ Civilian Fire Cas.-4 ☐ Apparatus-9
☐ Fire Serv. Casualty-5 ☐ Personnel-10
☐ EMS-6 ☐ Arson 11
☐ Rescue 12

J Property Use — Structures
Property Use
See back for "Common" Codes

K1 Person/Entity Involved — Local Option
Business name (if applicable) | Area Code Phone Number
☐ Check this box if same address as incident location. Then skip the three duplicate address lines.
Mr., Ms., Mrs. | First Name | MI | Last Name | Suffix
Number | Prefix | Street or Highway | Street Type | Suffix
Post Office Box | Apt./Suite/Room | City | State | Zip Code
☐ More people involved? Check this box and attach Supplemental Forms (NFIRS-1S) as necessary.

K2 Owner — ☐ Same as person involved? Then check this box and skip the rest of this section. — Local Option
Business name (if applicable) | Area Code Phone Number
☐ Check this box if same address as incident location. Then skip the three duplicate address lines.
Mr., Ms., Mrs. | First Name | MI | Last Name | Suffix
Number | Prefix | Street or Highway | Street Type | Suffix
Post Office Box | Apt./Suite/Room | City | State | Zip Code

M Authorization
Officer in charge ID | Signature | Position or rank | Assignment | Month Day Year
Check box if same as Officer in charge. ☐
Member making report ID | Signature | Position or rank | Assignment | Month Day Year

North Carolina Office of State Fire Marshal

Figure 11-3 Incident report.

Continued

Items With A ☆ Must ALWAYS Be Completed!

J	Property Use ☆ Structures		
131	Church, place of worship	341	Clinic, clinic type infirmary
161	Restaurant or cafeteria	342	Doctor/dentist office
162	Bar/tavern or night club	361	Prison or jail, not juvenile
213	Elementary school or kindergarten	419	1 - or 2- family dwelling
215	High school or junior high	429	Multi-family dwelling
241	College, adult ed.	439	Rooming/boarding house
311	Care facility for the aged	449	Commercial hotel or motel
331	Hospital	459	Residential, board and care
		464	Dormitory/barracks
		519	Food and beverage sales

539	Household goods, safes, repairs
579	Motor vehicle/boat sales/repair
571	Gas or service station
599	Business office
615	Electric generating plant
629	Laboratory/science lab
700	Manufacturing plant
819	Livestock/poultry storage (barn)
822	Non-residential parking garage
891	Warehouse

	Outside		
124	Playground or park	936	Vacant lot
655	Crops or orchard	938	Graded/cared for plot of land
669	Forest (timberland)	946	Lake, river, stream
807	Outdoor storage area	951	Railroad right of way
919	Dump or sanitary landfill	960	Other street
931	Open land or field	961	Highway/divided highway
		962	Residential street/driveway

981	Construction site
984	Industrial plant yard

L Remarks? Attach Supplemental Forms (NFIRS-1S) as necessary.

Fire Module Required?

Check the box that applies and then complete the additional Fire mod. Based on Incident Type as follows:

Buildings 111	Complete Fire & Structure
Special structure 112	Complete Fire Mod. & the I Block on Structure Module
Confined 113-118	Complete Basic Module
Mobile Property 120-123	Complete Fire Module
Vehicle 130-138	Complete Fire Module
Vegetation 140-143	Complete Fire or Wildland
Special outside fire 161-164	Complete Fire Module

Figure 11-3 Continued

CHAPTER QUIZ

1. Fires set to reduce or eliminate business competition are known as:
   a. direct gain fraud fires
   b. indirect gain fraud fires
   c. revenge fires
   d. spite fires
2. In testifying in court, it is important that your appearance and testimony be:
   a. judgmental
   b. opinionated
   c. professional
   d. reserved

3. Administrative search warrants are valid for the purpose of searching for:
    a. fire victims
    b. evidence of arson
    c. illegal drugs
    d. the cause of the fire

4. The reason for a fire is known as the:
    a. motive
    b. cause
    c. intent
    d. purpose

5. A reasonable cause of belief in the existence of facts is called:
    a. administrative search warrant
    b. a hunch
    c. probable cause
    d. suspicion

6. A first responder's opportunities for making observations occur:
    a. during initial dispatch
    b. while en route to the incident
    c. upon arrival at the scene
    d. all of the above

7. The origin of a fire is usually located at the:
    a. lowest level of burning
    b. highest level of burning
    c. top of the V pattern
    d. point where there is the least amount of fire damage

8. A legal document issued by a judge or magistrate that directs law enforcement officers to conduct a search and collect evidence is called:
    a. an administrative process
    b. an administrative search warrant
    c. probable cause
    d. a search warrant

9. A legal document issued by a judge or magistrate that directs law enforcement officers to conduct a search and provides details as to location and reason is called:
    a. an administrative process
    b. an administrative search warrant
    c. probable cause
    d. a search warrant

10. Immediate overhaul of the fire scene should be delayed if possible to:
    a. allow firefighters to rest
    b. allow the fire investigator to see if the arsonist will return
    c. allow time for the burn pattern to become more evident
    d. allow the investigator to determine the cause of the fire

CHAPTER 12

THE COMPANY OFFICER'S ROLE IN PLANNING AND READINESS

OUTLINE

- Objectives
- Overview
- What are Target Hazards?
- Pre-incident Planning
- What Should Pre-incident Plans Address?

- What Information Should Your Pre-incident Plan Include?
- Key Terms
- Review Questions

- Discussion Questions
- Homework
- Chapter Quiz

OBJECTIVES

Upon completion of this chapter, you should be able to describe:

- Pre-incident planning activities
- Training and education needs for company members
- The company officer's role in maintaining company readiness

OVERVIEW

Good pre-incident planning is essential for safe emergency scene management. The plan should provide information about building layout, access, construction features, occupancy, and a host of other important items that are needed before deciding to enter the building to attack the fire directly. The advantages of pre-incident planning are often overlooked. We continue to read of firefighters who are injured and killed in structural fire situations where there was no pre-incident plan, or, if there was a preplan, it was not followed. Pre-incident planning should address the very topics that may likely become an issue during any significant emergency in the building. These include layout, contents, construction, and type and location of installed fire protection equipment.

Company officers should develop a personal plan that will allow them to effectively lead their company to accomplish all of the tasks considered in this chapter. This will help the company become proficient in its tasks and ready to safely respond to citizens of the community it serves. The company's ability to provide these essential services lies with the knowledge and dedication of the company officer. Understanding problems and solutions will increase readiness and reduce challenges. Readiness also involves being ready to perform. Constant planning, training, and physical fitness are essential for company readiness (see **Figure 12-1** and **Figure 12-2**).

Figure 12-1 Preplanning information should be readily available for responding units.

Figure 12-2 Company personnel should work together to develop the information needed for the pre-incident planning process.

WHAT ARE TARGET HAZARDS?

Target hazards usually include:

1. Occupancies such as health care facilities, including convalescent homes.
2. Large public-assembly facilities, including airports, theaters, libraries, and so on.
3. Multiple residential occupancies, including apartment buildings, motels, and dormitories.
4. Schools and educational centers.
5. Occupancies such as jails and prisons.
6. Occupancies that present difficult challenges—highrise buildings, large buildings of any classification, and so on.
7. Occupancies where access by fire department vehicles precludes the normal method of operations.
8. Occupancies where there may be high property value.
9. Occupancies where there is a high likelihood of a fire.
10. Mercantile and business occupancies.
11. Industrial facilities and hazardous storage facilities.
12. Large unoccupied buildings.
13. Buildings under construction or demolition.

 # PRE-INCIDENT PLANNING

Pre-incident planning is a process of preparing a plan for emergency operations at a given building or hazard. Most pre-incident planning is directed to a specific location or occupancy. When the target of the planning process is a fixed location, the following information should be incorporated into the plan information:

1. Building number and street address
2. Other names commonly used to describe the occupancy
3. Occupancy load during the day and night
4. Building description (number of floors, dimensions, or square feet of floor space)
5. Hazards to personnel
6. Installed fire detection equipment
7. Installed fire suppression equipment
8. Water supply
9. Access to utility cutoffs
10. Priority salvage area
11. Special considerations
12. A basic floor plan (See **Figure 12-3**)

WHAT SHOULD PRE-INCIDENT PLANS ADDRESS?

1. Life hazards problem for occupants and firefighters
2. Available egress for the occupants
3. Availability of access for firefighters
4. Places where a fire would most likely start
5. Factors that would influence the spread of the fire

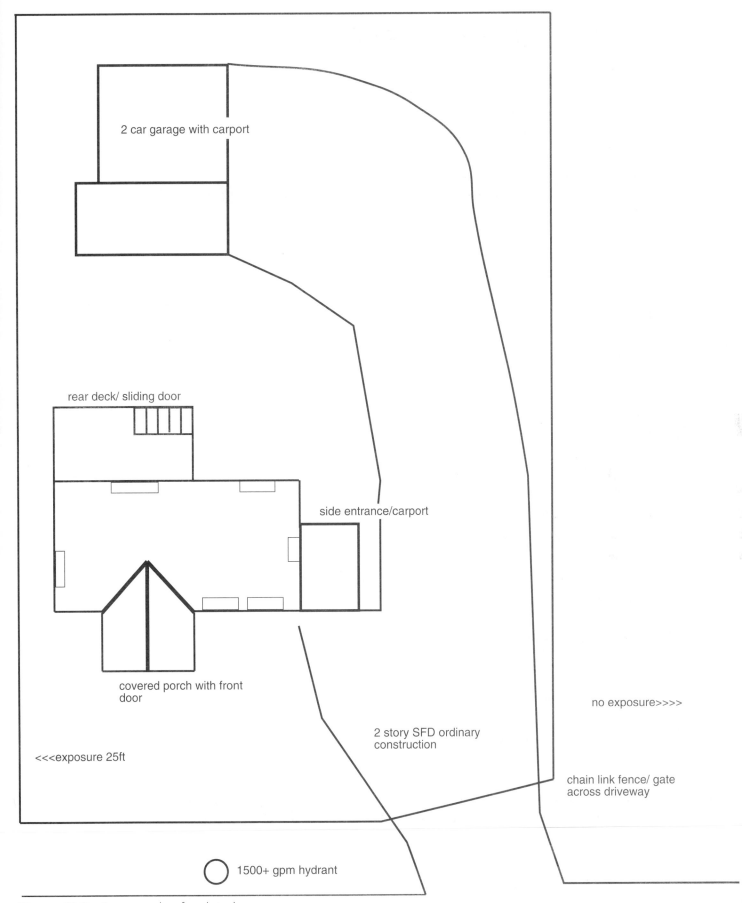

2 car garage with carport

rear deck/ sliding door

side entrance/carport

covered porch with front door

no exposure>>>>

<<<exposure 25ft

2 story SFD ordinary construction

chain link fence/ gate across driveway

1500+ gpm hydrant

Figure 12-3 An example of a plot plan.

6. Factors that would influence the intensity of the fire
7. Initial strategies for effective fire containment and control
8. Factors that would limit the fire department operations
9. Resource needs, including those of other agencies
10. Apparatus placement
11. Location of the command post

WHAT INFORMATION SHOULD YOUR PRE-INCIDENT PLAN INCLUDE?

1. Address
2. Owner
3. Occupancy
4. Means of access and entry
5. Personnel hazards
6. Fire behavior predictions
7. Locations of stairs and elevators
8. Ventilation systems
9. Installed fire protection systems
10. Exposures
11. Special hazards
12. Resource needs
13. Company assignments
14. Estimated fire flow
15. Water supply
16. Predicted strategies

Key Terms

Define in your own words the following terms:

area of refuge _____

_____

fire control _____

_____

floor plan _____

_____

line safety _____

_____

lock box _____

_____

plot plan _____

_____

pre-incident planning _____

_____

pre-incident survey _____

_____

property conservation _____

_____

Quick Action Prefire Plan _____

_____

SOP _____

_____

target hazards _____

_____

QUESTIONS

Review Questions

1. **a.** What is a pre-incident survey?

_____

_____

_____

_____

b. What is a pre-incident plan?

_____

_____

_____

_____

2. Why is it best to use fire suppression personnel to develop these plans?

_____

_____

_____

_____

3. What information should be contained in a pre-incident plan?

_____

_____

_____

_____

4. What features of the building should be noted in the pre-incident survey?

_____

_____

_____

_____

5. What types of occupancies should be preplanned?

_____

_____

_____

_____

6. How can the pre-incident planning information be made available during a time of need?

_____

_____

_____

_____

7. Should all members of the company be familiar with pre-incident plans for occupancies in their area? Why?

_____

_____

_____

_____

8. Why is it important to have a continuing program of company level training?

_____

_____

_____

9. What are the benefits of multiple-company training activities?

_____

_____

_____

_____

10. What is your role as the company officer in company readiness?

_____

_____

_____

_____

Discussion Questions

1. Identify several target hazards in your community. What makes these "target hazards"?

_____

_____

_____

_____

2. Take one of the occupancies you identified in the previous question and describe the process of gathering the information and developing a preplan for that situation.

_____

_____

_____

_____

3. How well do you preplan target hazards in your first-due response area?

_____

_____

_____

_____

4. What use do you make of these preplans?

_____

_____

_____

_____

5. How can the preplanning process be extended to emergencies other than fire (for example, EMS, hazardous materials release, or technical rescue)?

_____

_____

_____

_____

6. The company officer plays a significant role in the delivery of emergency services. To be effective he or she must be able to deliver these services by solving problems in an effective and efficient manner. How can company officers improve their company readiness in each of the following areas:

a. Anticipation

_____

_____

_____

_____

b. Preparation

_____

_____

_____

_____

c. Action

_____

_____

_____

_____

7. Why is time so critical in response to a fire or EMS event?

_____

_____

_____

_____

8. FEMA's programs are focused on mitigation, preparedness, response, and recovery. Given the type of natural disaster most likely in your area, how should local, state, and federal resources prepare for and respond to such a disaster?

_____

_____

_____

_____

9. How well prepared is your fire department to deal with:

   a. A structural collapse with fifty or more injured occupants

   _____

   _____

   _____

   _____

   b. A HAZMAT event involving a toxic chemical and ten or more affected citizens

   _____

   _____

   _____

   _____

   c. A response to an event such as you proposed in question 8

   _____

   _____

   _____

   _____

10. How does your company's training program measure up in preparing firefighters for emergency response?

   _____

   _____

   _____

   _____

HOMEWORK FOR CHAPTER 12

For Fire Officer I

What NFPA 1021 says:

This duty involves supervising emergency operations, conducting preincident planning, and deploying assigned resources in accordance with the local emergency plan and according to the following job performance requirements.

The Fire Officer I shall "develop a preincident plan, given an assigned facility and preplanning policies, procedures, and forms, so that all required elements are identified and the approved forms are completed and processed in accordance with policies and procedures."

Situation:

If, as described in the text, life safety is our greatest concern, then our list of target hazards should include facilities where life safety is our major concern, places like hospitals, assembly occupancies, nursing homes, and similar assisted-care residential facilities.

Given that the likelihood of fire or similar emergency in an assembly occupancy or institutional facility such as a hospital is relatively low, let us focus our preincident planning energies on residential facilities where older people may be living. Preplanning these facilities is very appropriate since over the past three years there have been several fires in facilities of this type, often with tragic results. The people who live there tend to be older and may have some medical condition that would present a challenge to their self-preservation during an emergency. Some of these facilities have reduced staffing at night, and some are in structures that are not the safest.

Your assignment: Following the SOPs of your department, and using the forms provided by your department, conduct a preplan of a residential care facility in your first-due area. This assignment includes the actual process of visiting the site, gathering the information, and preparing the preincident planning document. (Do not submit an existing preincident plan.)

When you have completed the assignment, keep the original documents for future use by your fire department, and provide a copy of the documents to your instructor.

See the textbook for forms that will assist your efforts. Make copies if needed to complete and include with your preincident plan.

For Fire Officer II

What NFPA 1021 says: This duty involves supervising multi-unit emergency operations, conducting preincident planning, and deploying assigned resources, according to the following job requirements.

The Fire Officer II shall "produce operational plans, given an emergency incident requiring multi-unit operations, so that required resources and their assignments are obtained and plans are carried out in compliance with approved safety procedures resulting in the mitigation of the incident."

Situation:

In this situation, the occupancy is the highway and the event is a multi-vehicle accident involving several injured people. Many departments have prepared for such events with an SOP that mobilizes needed resources for the initial response, rather than calling each agency and unit as needed. The fire department will normally respond to handle extrication and any fires that may result from the event. EMS, as a separate agency or as part of the fire department, should provide at least two units to the scene. Appropriate police agencies should respond to help manage traffic and provide the necessary administrative functions at the scene of the accident. You may want to simply assign a unit to the lanes and protect the workers, as described in Chapter 8.

While we have not yet looked at incident management in detail, you may want to look at Appendix C of the textbook to see if there are any applicable provisions that should be considered.

Your assignment: Develop an SOP for the situation described above and submit it to your instructor. The SOP should address, as a minimum, resources necessary to mitigate the situation, and the command structure that would operate at such an event. Assume that this operation will allow rescue to be accomplished, the injured to be transported, and the highway to be cleared in one hour or less.

CHAPTER QUIZ

1. The first priority during emergency operations should be focused on:
   a. fire control
   b. incident stabilization
   c. life safety
   d. property conservation

2. The effort to reduce primary and collateral damage is called:
   a. fire control
   b. incident stabilization
   c. property conservation
   d. life safety

3. A bird's eye view of a facility showing structures, access points, and water supply is called a:
   a. plot plan
   b. floor plan
   c. blueprint
   d. preplan

4. Pre-incident planning includes:
   a. the pre-incident survey
   b. the development information resources
   c. the development of procedures to be used during emergency situations
   d. all of the above

5. Typical target hazards include:
   a. hospitals
   b. schools
   c. unoccupied buildings
   d. all of the above

6. The fact-finding portion of the emergency planning process is called the:
   a. assessment
   b. plan
   c. study
   d. survey

7. The first priority of the facility survey is to identify:
   a. decisions regarding the mode of attack
   b. installed fire protection features
   c. resource needs
   d. the attitude of the owner

8. The responsibility for maintaining readiness rests with the:
   a. company officer
   b. fire chief
   c. individual firefighters
   d. training officer

9. Multi-company training activities permit firefighters to:
    a. compete with other companies
    b. see how things will work during real events
    c. see old friends
    d. all of the above

10. Officers should continue their personal training by:
    a. attending seminars
    b. enrolling in college
    c. reading the trade journals
    d. all of the above

THE COMPANY OFFICER'S ROLE IN INCIDENT MANAGEMENT

OUTLINE

- Objectives
- Overview
- Size-Up Factors
- Incident Priorities and Fireground Tasks

- Key Terms
- Review Questions
- Discussion Questions

- Homework
- Chapter Quiz

OBJECTIVES

Upon completion of this chapter, you should be able to describe:

- The factors to be considered during size-up
- The factors that determine how fire spreads in a structure
- The procedures to control, confine, and extinguish fires and protect exposures in structure fires, outdoor situations, and where hazardous materials may be present
- The duties and responsibilities of officers using the incident management system at responses involving one or more units
- The legal issues for fire officers

OVERVIEW

Company officers should be able to analyze emergency scene conditions, conduct size-up, develop and implement an initial action plan, and deploy resources to safely and effectively control an emergency.

If the event requires more than one or two companies, some form of Incident Management System should be in place. As a company officer you may be in charge of your company or you may be given an assignment as part of the management team. In all cases, you should be familiar with the basics of the Incident Management System (see **Figure 13-1**).

While the focus of this chapter and the entire text is to help you satisfy the performance standards of NFPA 1021, company officers should be able to cope with routine EMS and hazardous materials events as well. Today, most firefighters are certified as EMTs and hazardous materials responders at the operations level. Training for both programs should include managing incidents in these specialty areas. If you can integrate the IMS

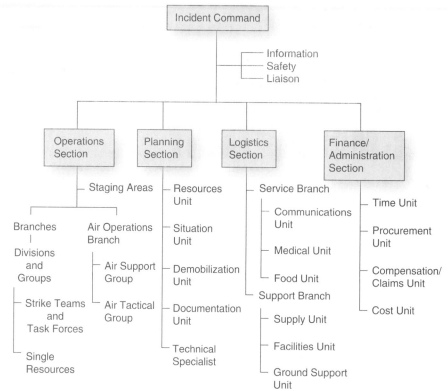

Figure 13-1 Incident Management System.

information from the text with what you have learned in those courses, you should be ready to manage routine events involving medical emergencies, hazardous materials incidents, and even kitchen fires.

SIZE-UP FACTORS

There are several acronyms that can be used for keeping track of the size-up factors. One of these is **COAL WAS WEALTH.**

Construction	**W**ater	**W**eather
Occupancy	**A**uxiliary appliances	**E**xposures
Apparatus and personnel	**S**treet conditions	**A**rea
Life safety		**L**ocation
		Time
		Height

Another acronym used to list size-up factors is **WALLACE WAS HOT.**

Weather	**W**ater	**H**eight
Apparatus (and staffing)	**A**uxiliary equipment	**O**ccupancy
Life hazard	**S**pecial hazards	**T**ime
Location and extent of fire		
Area involved		
Construction		
Exposures		

Keeping track of this information can be quite a challenge. Many departments use a tactical worksheet for larger events, but there are benefits in keeping some sort of check-off list for *all* events. Many officers keep a worksheet

POST INCIDENT ANALYSIS

SUMMARY

DATE _____   TIME OF ALARM _____

ADDRESS _____

TYPE OF INCIDENT _____

SITUATION UPON ARRIVAL OF FIRST UNITS: (INCLUDE A BRIEF DESCRIPTION OF THE SITUATION ENCOUNTERED BY THE FIRST UNIT(S) ARRIVING ON THE SCENE. THE TYPE OF UNITS AND MANPOWER ON UNITS SHOULD BE LISTED.)

_____
_____
_____
_____
_____
_____
_____
_____
_____
_____

FINAL OUTCOME OF INCIDENT: (LIST EXTENT OF DAMAGE AND CASUALTIES. ALSO INCLUDE DAMAGE TO FIRE EQUIPMENT AND EMERGENCY PERSONNEL CASUALTIES.)

_____
_____
_____
_____

EQUIPMENT COMMITTED TO INCIDENT: (LIST MANPOWER AND UNITS COMMITTED TO THE INCIDENT. INCLUDE PARTICIPATING VOLUNTEERS AND PAID OFF DUTY PERSONNEL THAT RESPONDED IN PRIVATE AUTOMOBILES.)

_____
_____
_____
_____
_____
_____

Figure 13-2 Every event should be followed by an analysis of what happened. A simple form encourages participation and channels ideas along productive paths.

Continued

folded up in the pocket of their turnout coat. When the event is all over, take time to analyze your actions and learn from your experiences. Forms may help you in this process. A sample postanalysis form and tactical worksheet are shown on the following pages (see **Figures 13-2** and **13-3**). These sheets provide check-off lists for size-up

EQUIPMENT AND MANPOWER NOT COMMITTED TO THE INCIDENT: (LIST THE STATIONS LEFT EMPTY AND THOSE STATIONS BACKFILLED BY WHAT APPARATUS.)

_____

_____

_____

STRATEGY: (LIST THE INCIDENT COMMAND STRATEGIES CHOSEN. INCIDENT COMMANDERS SHOULD DESCRIBE THEIR BASIC PLAN TO ADDRESS THE PRIORITIES OF THE INCIDENT AT THE TIME THEY BECAME THE INCIDENT COMMANDER.)

FIRST IN UNIT(S): _____

A. GENERAL STRATEGY _____

_____

_____

_____

B. RESULTS _____

_____

_____

_____

FIRST INCIDENT COMMANDER (NAME): _____

A. GENERAL STRATEGY _____

_____

_____

_____

B. RESULTS _____

_____

_____

_____

SECOND INCIDENT COMMANDER (NAME): _____

A. GENERAL STRATEGY _____

_____

_____

_____

B. RESULTS _____

_____

_____

Figure 13-2 Continued

factors, a place to record assignments, a space for a simple diagram, and spaces to describe incident analysis and strategies.

Size-up must be an ongoing process that continues throughout the event. Many responses start without all the information, or firefighters find that the information provided is incorrect. As additional information is obtained, be sure it is shared with those who need it. Everyone involved in the incident should share and benefit from the process.

Sample 1

Tactical Worksheet

Address: _____

Occupancy: _____

Incident No.

Time

Wind Direction	Personnel Accountability (PAR)	Tactical	Benchmark	Functional

Wind Direction
_____

Elapsed Time
5 10 15 20 25 30 PAR

Level II Staging
_____

Personnel Accountability (PAR)

All Clear

30 Min.

Under Control

Off-To-Def

Hazardous Event

No "PAR" Upgrade Assign.

Tactical
- Overall Plan
- Water Supply
- Search & Rescue
- Initial Attack
- Exposures
- Rapid Intervention Team
- Logistical Needs
- Ventilation
- Evacuation
- All Clear
- Fire Control
- Salvage (Loss Stopped)
- Accountability

Benchmark   Functional
- Command Location
- Pumped Water
- Gas
- Electrical
- Recon
- Outside Agency
- Investigator
- P.P.V.
- P.D.
- Primary -
- Secondary
- Salvage (Loss Stopped)
- C.O. Meter

E
E
E
E
L
L
H
R
U
BC

E
E
E
E
L
L
H
R
U
BC

Branch   Command   Branch

Figure 13-3   A sample tactical worksheet.

INCIDENT PRIORITIES AND FIREGROUND TASKS

Incident priorities are:

1. Life safety
2. Incident stabilization
3. Property conservation

Lloyd Layman's Tactics start with size-up. He lists five activities that are standard events in controlling fires. The two supporting activities may be required at any time.

ALWAYS	**ESSENTIAL TASKS**	**SUPPORTING TASKS**
size-up	rescue	ventilation
	exposures	salvage
	confinement	
	extinguishment	
	overhaul	

Key Terms

Define in your own words the following terms:

action plan _____

_____

attack role _____

_____

benchmark _____

_____

command role _____

_____

defensive mode _____

_____

environmental factors _____

_____

fire confinement _____

_____

fire control _____

_____

fire extinguishment _____

_____

initial report _____

_____

mutual aid _____

_____

offensive mode _____

_____

overhaul _____

_____

strategy _____

_____

tactics _____

_____

tasks _____

_____

transition mode _____

_____

QUESTIONS

Review Questions

1. What are incident priorities?

_____

_____

_____

_____

2. What is meant by command sequence?

_____

_____

_____

_____

3. What is meant by size-up?

_____

_____

_____

_____

4. What are the parts of size-up?

_____

_____

_____

_____

5. What are the three modes of firefighting?

_____

_____

_____

_____

6. What information should be included in the initial report?

_____

_____

_____

_____

7. What is an action plan?

_____

_____

_____

_____

8. What do the terms _strategy_ and _tactics_ mean?

_____

_____

_____

_____

9. What is IMS? When should it be used?

_____

_____

_____

_____

10. What is your role as company officer in incident management during the early phases of fires and other emergency activity?

_____

_____

_____

_____

Discussion Questions

The first six questions pertain to the kitchen fire in Ms. Garcia's home.
1. What were the size-up factors at Ms. Garcia's house?

_____

_____

_____

_____

2. As the first arriving officer at this fire, what would be your size-up report?

_____

_____

_____

_____

3. Considering Chief Layman's list of strategies, which would be relevant at this fire?

_____

_____

_____

_____

4. What would you say to Ms. Garcia?

_____

_____

_____

_____

5. What would you do for her?

_____

_____

_____

_____

6. What could you do to protect the rest of her home while you extinguished the kitchen fire?

_____

_____

_____

_____

7. How would you use IMS at the scene of a traffic accident on an interstate highway?

_____

_____

_____

_____

8. Using IMS principles, describe the implementation of strategy and tactics required for controlling a fire at each of the following:

a. Structure

_____

_____

_____

_____

b. Flammable spill fire following an accident

_____

_____

_____

_____

c. Hazardous materials release from a railroad car in an urban area

_____

_____

_____

_____

d. Fire in the wildland/urban interface

_____

_____

_____

_____

9. Are there any unoccupied buildings in your jurisdiction? If so, what is your department doing in anticipation of a response to a fire or other event in one of these buildings? What preparation is being taken at the company level, and what action would be appropriate if you were called to a structural fire in one of these buildings today?

_____

_____

_____

_____

10. What is the two-in two-out rule? How does it affect operations in your jurisdiction?

_____

_____

_____

_____

HOMEWORK FOR CHAPTER 13

Emergency Service Delivery for Fire Officer I

What NFPA 1021 says:

Section 4.6 "This duty involves supervising emergency operations, conducting preincident planning, and deploying assigned resources in accordance with the local emergency plan and according to the following job performance requirements."

  a. specifically section 4.6.2 says that the Fire Officer I shall: "Develop an initial action plan, given size-up information for an incident and assigned emergency response resources, so that resources are deployed to control the emergency.

  b. section 4.6.3 says that the Fire Officer I shall "Implement an action plan at an emergency operation, given assigned resources, type of incident, and a preliminary plan, so that resources are deployed to mitigate the situation.

  c. and section 4.6.4 says that the Fire Officer I shall "Develop and conduct a postincident analysis, given single unit incident and postincident analysis policies, procedures, and form, so that all required critical elements are identified and communicated, and the approved forms are completed and processed in accordance with policies and procedures."

Situation for parts a and b:

FIRE AT THE LAWN SHOP

It is 9:00 P.M. on a Saturday night and the weather is the same as it is today. You are the officer on the first arriving unit to a fire-alarm activation in a small neighborhood store that sells various agricultural products to homeowners for their plants and lawns (see **Figure 13-4**). Upon arrival you find there is a fire in the office area of the building. You ask for a full alarm assignment. Eventually the personnel of three engine companies, one truck company, two battalion chiefs, and an EMS unit are committed to the event.

Fortunately, water is not a problem and your troops are able to establish a reliable and adequate water supply. However, you are concerned about any reaction of the products involved to water, as well as the contamination of any water used for firefighting.

Because of the nature of the event the regional HAZMAT team is called in to provide some help and direction in dealing with any hazardous materials that may be present. Since it was a "slow news" period, members of the press and spectators of all sorts join you. Lots of folks are driving by and stopping to see what you are doing.

Eventually you are relieved by one of the battalion chiefs and are detailed to head the operations sector for fire suppression.

Your assignment: Develop and describe an initial action plan, given the size-up information for an incident and assigned emergency response resources, so that resources are deployed to control the emergency.

Figure 13-4 This neighborhood garden shop is the setting for the homework for this chapter.

As the initial arriving officer, you are the incident manager until relieved by the battalion chief. Describe how you would implement your action plan, given the assigned resources, type of incident, and a preliminary plan, so that resources are deployed to mitigate the situation.

Explain and diagram (via organization chart) the incident management structure for your relief.

Copy and complete the Tactical worksheet, Figure 13-3, or use a similar form from your department. Complete the form for as much information as you would have had and for the action you would have taken prior to your relief.

Incorporate into your response the requirements of any applicable SOPs.

Situation for part c: The activity requires a post-incident analysis. There are several options that will satisfy this requirement.

First, the instructor may ask that you comment on the work of a classmate.

Second you may be asked to analyze an event that recently occurred in your community. At the Fire Officer I level, this need not be anything more complicated than a single unit response. The intent is to get you to recognize the benefit of such an analysis after each event.

Your assignment: Conduct a post-incident analysis using the information and forms provided such as Figure 3-2.

For Fire Officer II

Situation:

Again, the example used for Fire Officer I may be discussed. Second, you may be asked to analyze an event that recently occurred in your community.

As an alternative you may be asked to analyze an event that recently occurred in your community. At the Fire Officer II level, this should be an event that required a multi-unit response. It need not be a fire, and in fact, looking for something unusual may provide greater interest and benefit. Again, the intent is to get you to recognize the benefit of such an analysis after such an event.

Your assignment: Conduct a post-incident analysis using the information and forms provided.

CHAPTER QUIZ

1. Incident priorities are:
   a. thinking, planning, and acting
   b. size-up, strategy, and tactics
   c. life safety, incident stabilization, and property conservation
   d. all of the above

2. When transmitting the accomplishment of the initial search phase of the operation, the message should be:
   a. all clear
   b. under control
   c. loss stopped
   d. a-ok

3. Size-up is an information-gathering step that is:
   a. ongoing with constant observation
   b. only used during preplanning activities
   c. started after the strategic objectives have been met
   d. started at the scene of every incident

4. Communications between the incident command and other units should be:
   a. brief and allow for interpretation by unit personnel
   b. clear and concise with a confirmation of understanding
   c. held to a minimum
   d. long and detailed as needed

5. The term _____ identifies the specific use of a building.

    a. classification

    b. fire load

    c. occupancy

    d. ownership

6. The _____ report should paint a concise but vivid picture of the conditions at the scene, as well as a quick summary of your intentions and needs.

    a. brief

    b. initial

    c. oral

    d. written

7. Which of the following is not an operational strategy?

    a. advancing hand lines

    b. rescue

    c. salvage

    d. ventilation

8. Effective fireground communications:

    a. improve firefighter safety

    b. improve interagency cooperation

    c. increase effectiveness

    d. all of the above

9. Strategy is:

    a. the overall plan to control the incident

    b. the operations that need to be done

    c. the method of implementing an action plan

    d. always the same

10. The incident management system should be used:

    a. upon arrival of the first battalion chief

    b. upon arrival of the first unit

    c. when directed by dispatch

    d. only at "the big one"

ACRONYMS

ADA	Americans with Disabilities Act
BOCA	Building Officials and Code Administrators
BTU	British thermal unit
CIS	Critical incident stress
CPSC	Consumer Product Safety Commission
DHS	Department of Homeland Security
DOE	Department of Energy
EEOC	Equal Employment Opportunity Commission
EMS	Emergency medical services
EMT	Emergency medical technician
HSO	Health and safety officer
IAFF	International Association of Fire Fighters
IDLH	Immediately dangerous to life and health
IMS	Incident management system
JPR	Job performance requirement
MBE	Management by exception
MBO	Management by objectives
NFA	National Fire Academy
NFIRS	National Fire Incident Reporting System
NFPA	National Fire Protection Association
NIH	National Institutes of Health
OSHA	Occupational Safety and Health Administration
PASS	Personal alert safety system
PCFD	Plain City Fire Department
PPE	Personal protective equipment
PPV	Positive pressure ventilation
SCBA	Self-contained breathing apparatus
SOG	Standard Operating Guideline
SOP	Standard operating procedure
TQM	Total quality management
USFA	United States Fire Administration

"Hey, let's be careful out there!"